SpringerBriefs in Evolutionary Biology

For further volumes:
http://www.springer.com/series/10207

Bernard Marcus

Evolution That Anyone Can Understand

Dr. Bernard Marcus
Genesee Community College
Batavia, NY 14020
USA

ISSN 2192-8134 e-ISSN 2192-8142
ISBN 978-1-4419-6125-9 e-ISBN 978-1-4419-6126-6
DOI 10.1007/978-1-4419-6126-6
Springer New York Dordrecht Heidelberg London

Library of Congress Control Number: 2011938146

Printed on acid-free paper

Springer is part of Springer Science+Business Media (www.springer.com)

To Gracie and Zoë

Preface

When I first mentioned to a colleague that I wanted to write a book on evolution, he asked me why another one was necessary; were not there already plenty in print? Indeed there were, many of them better than anything I could hope to write. But most with which I was acquainted were not written for the audience I had in mind. I wanted to write a book that someone with no more than a course in high school biology could pick up and read. Among those, I hope, will be some who do not accept the reality of evolution.

I do not presuppose that my writing is so clear and persuasive that anyone who reads it will be magically convinced that evolution occurred. Rather, I would hope that he or she would use this book as a springboard to try to learn more about the subject. So many people who reject evolution know so little about it, they have no idea how hollow their arguments against it happen to be.

However, my more sincere hope is that someone who is unacquainted with but open-minded toward the subject will also use this as a first step toward learning more. In hope of facilitating that, I have included some selected references in each chapter. I have deliberately not tried to document every point I have made because I believe that many whom I want to read this book would be, to use a cliché, turned off by long lists of appended endnotes. Rather, I added a few that I thought would be both appropriate and understandable. In addition, for a few points I have added references from the scientific literature for those who might want to tackle them. For the most part, however I have avoided those.

Finally, I wish to acknowledge that there are some evolution scientists who have written excellent popular books and articles about the subject. I do not count myself among them. But I would hope that anyone who picks up this book would also read the late Stephen Jay Gould, Sean Carroll, Richard Dawkins, and other scholars who have made the information available to anyone who wants to take advantage of it.

Rochester, NY Bernard A. Marcus

Introduction

The function of scientific research is to promote the understanding of the world around us. In theory, anyway, the more we learn, the more potential we have of making our lives better. Thus, we have seen that research in electronics provides us with computers, research in chemistry provides us with all manner of synthetics, and research in agriculture provides us with more food. Periodically, scientific research uncovers something that makes some of us uncomfortable. The discovery of the link between smoking and lung cancer and heart disease was not received well by the tobacco industry, and the link between global climate change and fossil fuel use has not been well received by the petroleum industry, to cite just two examples. Usually the response of those whose world has been disrupted by science is denial, often followed by attack on or ridicule of the science that has challenged them. In the long term, however, science usually turns out to be correct.

For more than the past 100 years, such a battle has been going on between some members of the religious community and one branch of science: evolutionary biology. The discoveries in evolution are dogmatically denied by the fundamentally religious. One can almost imagine similar arguments between monotheists and polytheists two millennia ago, but in this case, the battle is between fact and belief, and it is completely void. Acceptance of one side does not categorically reject the other.

Part of the reason for the disagreement is that many people who reject evolution know nothing about it. Their thesis is based on misinformation and misunderstanding. Indeed, that may be true in the case of some of those who accept it as well. In this book, I am going to attempt to describe some of the evidence behind evolution. My goal is not to win converts; it is simply to provide information to those who would like to learn more about subject, even those who, in the end, will continue to reject evolution. I hope that they will at least have gained insight into the theory.

Contents

Chapter 1
Exactly What Is Evolution?

Nobody knows what time of day it arrived nor from where it came, other than somewhere in outer space. After all, there was nobody here to see it. Perhaps some of the animals looked skyward and noticed it. Some may have fled in fear, finding a convenient hole to hide in or tree to hunker under. Some, depending on what kind of intelligence they may have had, might even have found themselves bewildered, having seen nothing like it before. Comet, planetoid, meteor? The words did not exist then, but some astral body figured to be 10 km—6 miles—across, roughly 65 million years ago, is thought by many scientists to have streaked over the sky and struck the earth in the shallow ocean near what is now the Yucatan Peninsula of Mexico, pulverizing itself in the gypsum substrate. One can only imagine the result. There must have been massive tsunamis at the very least, and tons of debris from the crushed comet and the ocean floor. Water from the impact in combination with all of the solid particulates undoubtedly erupted into the atmosphere, filling it with a thick aerosol that blocked the sun. Fragments of the comet, glowing hot from friction with the air, may have fallen onto land touching off massive firestorms. Countless numbers of plants and animals must have died as a direct result of the impact, but that was only the beginning. As impact debris rose into and spread through the upper atmosphere, absorbing and blocking sunlight, night must have come quickly, with temperatures dropping precipitously. Robbed of light, plants ceased their photosynthesis and quickly died. Plant eating animals may have survived for a while, eating whatever plant tissue was available to them, and animal eaters probably survived longer, feeding mostly as scavengers on the remains of whatever dead prey they could find. In the long run, however, they too died out. No doubt the gypsum that was blown into the air greatly acidified the rain that eventually began to wash it back to earth. Gypsum, largely made up of calcium sulfate, would react with atmospheric water vapor to form sulfuric acid. The disturbance to the planet's life-support system may have gone on for a few years, but the massive death it dealt was probably over within a matter of weeks, possibly a few months, a nanosecond in comparison to the 3.5 billion years or more that life has existed.

B. Marcus, *Evolution That Anyone Can Understand*,
SpringerBriefs in Evolutionary Biology, DOI: 10.1007/978-1-4419-6126-6_1,

The comet, or whatever it was, marked the end of the age of dinosaurs on Earth. Estimates have it that the comet wiped out 85% of all species on Earth at that time, but for the dinosaurs it is thought by most scientists to have been total annihilation. Those few dinosaur fossils that have been found in post-comet sediments are thought to have been deposited there after being eroded from their original locations.

Dinosaurs thrived on Earth for most of the Mesozoic Era, a span of geologic history that lasted for almost 150 million years, from about 213 million years ago to about 65. The Mesozoic is divided into three periods: the Triassic, during which the dinosaurs originated, the Jurassic, when birds originated and dinosaurs underwent a massive diversification, and the Cretaceous, when dinosaurs underwent another diversification. Mammals may have originated during the Jurassic, and they coexisted with dinosaurs throughout the cretaceous. The first mammals were small, inconspicuous, perhaps mouse-like in appearance, probably nocturnal and most likely elusive, given with whom they shared the planet. They most likely existed the way comparable mammals exist now, by feeding on seeds and insects and by burrowing.

The Cretaceous period and Mesozoic era ended with the comet, yielding to the Cenozoic Era, the so-called age of mammals. Mammals were able to survive the comet because they can maintain a constant internal body temperature, often described as being warm blooded, and they could conceivably huddle together in burrows to share body heat. Seeds were still available to those that could take advantage of them, and it is even conceivable that they could have scavenged among the corpses of the animals that had died. Moreover they were small; they could squeeze into tight places where they were protected from the harsh conditions that existed following impact.

Once enough debris had settled out of the atmosphere to allow penetration of sunlight, temperatures warmed and seeds that had formed and gone dormant before the event and had avoided being eaten by primitive mammals were able to sprout. With no competition from standing plants, they most likely prospered and began spreading, just as weeds today quickly overtake a vacant field. Mammals too, again with no competition though perhaps more importantly with no predators, were able to prosper and spread. In time, both plants and animals diversified to fill vacant niches and, over the 65 million years since the comet, gave rise to the organisms of today, ourselves included.

The preceding scenario introduces a number of ideas that are involved in the history of life and the history of our planet. Some would say that the ideas are controversial and that there is much contention surrounding them. In fact, the contention, controversy, disagreement, or whatever term one would choose to apply is overstated. Among the scientists who study this topic, there is remarkably little disparity; whatever differences exist among them is largely limited to detail. Virtually all agree in concept. The major deviation in thought comes mostly from the religious community, and only one part of that, really. Politically, this group has been referred to as the religious right, and they have been fighting to force their view on the rest of us. Others accept evolution but are still misinformed about it.

In the pages that follow, I will be discussing the subject in an attempt to add some clarity. At the very least, I hope I do not add to the confusion.

The major idea behind evolution is that life changes in form and function over time. The British naturalist Charles Darwin, who lived between 1809 and 1882, is usually credited with having framed the theory, but there were others who contributed to it. Darwin's grandfather, Erasmus Darwin, for example, framed a theory of evolution in 1794, as did Jean-Baptiste Lamarck in 1809. Lamarck's idea, that acquired traits were inherited, was incorrect, of course, but he still recognized that species of living organisms change over time. Darwin's insight into the process, what he called natural selection, is what made him unique. His epiphany reportedly occurred during a round-the-world voyage on the HMS Beagle, a mapmaking expedition of the British Navy. The voyage lasted from 1831 to 1836, during which Darwin spent much time exploring wherever the ship put into port. A good deal of this was in South America where he is said to have made two keen observations. These were that animals tend to bear more young than can possibly survive and that the young generally differ from their siblings in one or more ways. From the first observation, he concluded that individuals *struggle* (italics mine) to survive, and from the second that only those with the best characteristics or traits for their particular circumstances succeed. It is the environment or nature that determines or selects what trait or set of traits renders an individual fit to endure; those who lack the favorable characteristics either perish or fail to reproduce, thus giving only the best adapted individuals, often described as the most fit, the opportunity to perpetuate their traits in subsequent generations. This has come down to us today in the slogan "survival of the fittest," which was coined not by Darwin but by Herbert Spencer, an English philosopher, in 1864. Darwin further noted that a given animal species was likely to vary in appearance from place to place, indicating that different traits are advantageous and different selective pressures were at work in different locations.

The idea of animals varying in space is plainly obvious on the Galapagos Islands, an equatorial archipelago in the Pacific Ocean a bit less than 1,000 km (roughly 510 miles) off the coast of Ecuador in South America. Common knowledge has it the Darwin reached his conclusion after visiting them, but some biologists argue that what he saw there merely confirmed an idea that was already forming in his mind. The islands are home to a group of finches, now known appropriately enough as Darwin's finches. There are 13 species, all finches though not all particularly resembling one another. The island species vary mostly in terms of the structure of their beaks and their lifestyles or, more appropriately, niches, and each beak type appears to be adapted to a specific niche. For example, the so-called Large Ground-finch (*Geospiza magnirostris*), typically feeds on large nuts and seeds, which it cracks in its beak much as we would use our teeth to crack a peanut shell. Its beak is robust. In contrast, the Medium Ground-finch, *Geospiza fortis*, feeds on smaller seeds and has a correspondingly less robust beak. Another, the Vampire Finch, *Geospiza difficilis,* has a long, sharp beak. It has been known to probe flowers for nectar, but it also occasionally feeds on the blood of seabirds, which it obtains by pecking at its prey's skin.

Another animal Darwin observed on the Galapagos was the giant tortoise *Geochelone elephantopus*. These reptiles weigh as much as 250 kg (550 lbs) and are often greater than a meter (up to 4 feet) in length. They belong to a single species, but each of 14 islands has or had its own subspecies. In fact, it has been said that anyone who knows the tortoises well enough can be blindfolded and put on an island and know where he is as soon as he spots a tortoise.

Darwin also studied geology and was well acquainted with fossils. He was aware that these were the remnants of once living organisms, and so he was able to understand that animals vary in time as well as in space. Thus he came to appreciate that selective pressures may change over time as well and that a trait that is advantageous at one point may not be so at a later point. Consequently, with different selective pressures at work, a given species of organism may have a different appearance than a direct ancestor, something Darwin described as descent with modification.

Darwin returned to England from his voyage in 1836, but he did not publish his ideas on evolution until 1859. It is probable that he took some time after his return to study specimens he collected and crystallize his ideas, but it is also likely that he was apprehensive about the controversy he would cause. He did, however, continue to study and write in natural history, and he made a name for himself as a reputable scientist.

Darwin was prompted to finally publish by the work of another British naturalist, Alfred Russel Wallace (1823–1913). Working independently of Darwin first in South America and then in Southeast Asia, Wallace came to a conclusion that, while not identical to Darwin's, was close enough to qualify him too as a pioneer in evolution theory. Wallace had been in communication with Darwin; he sent Darwin an essay about his perception of evolution, and, realizing that he could no longer sit on the idea, Darwin published his ideas listing Wallace as co-author.

There were other British naturalists who contributed to the theory: geologist Charles Lyell (1797–1875), Robert Chambers (1802–1871), and Henry Walter (HW) Bates (1825–1892) to name just three. Other scientists like Thomas Henry Huxley (1825–1895) helped disseminate the theory. The point to be taken is that the theory of evolution developed as a result of a number of competent and accomplished scientists who came to their conclusions after much work, thought, and discussion. It was not a conclusion that was jumped to by a single person. Darwin gets much of the credit. Perhaps he deserves it, perhaps not. But evolution, like virtually all scientific advances, came about as a result of the efforts of many contributors.

Darwin and his colleagues were aware that all organisms have inheritable traits, but they were unaware of how these traits are controlled or passed on to succeeding generations. The former question has been answered by discoveries in molecular biology, particularly the functioning of DNA. This has occurred over the last 50–60 years, well after the deaths of the pioneering evolutionists. The matter of inheritance was actually answered during the lifetimes of those men mentioned in the previous paragraph, though they were not aware of it. It was accomplished by an Austrian monk by the name of Gregor Mendel (1822–1884). Mendel worked with garden peas crossing them for different traits out of apparent curiosity. He discovered, among other things that when two plants with dissimilar expressions for a given trait were crossed, one expression often appeared in the next generation while the other did not. The recessive trait, the one not appearing, however, was not lost; it could be retrieved in a successive generation. The factors controlling traits are, of course, the genes. Unfortunately, as the story goes, Mendel published his findings in an obscure journal; his contemporaries remained ignorant of them. As will be described in Chap. 7, that story may be apocryphal. Since Mendel's time, however, it has been learned that genes control virtually all of our physical traits and many of our psychological ones as well. Moreover, the interaction between genes is often more complicated than one trait simply being expressed and the other being hidden. Sometimes there is a blending of the two with the offspring appearing to be intermediate between their parents, and sometimes both are expressed, with the offspring appearing to be a mosaic of the parents. Furthermore, many traits are not controlled by a single gene or gene pair; many are controlled by multiple pairs of genes in combination. Sometimes, gene combinations can be debilitating or lethal, such as traits like albinism in plants or hemophilia or Duchene's muscular dystrophy in humans. Again, this is discussed more fully in Chap. 7. Darwin's observation of animals varying in space is explained by different gene combinations in those animals being more fit in, that is better adapted to, one environment than another. Thus, for example, darker coloration that allows a warm-blooded animal to radiate away excess heat would be advantageous in a hot climate, while light coloration that tends to conserve heat would be better for the same species of animal in a cold climate. By the same token, if selective pressures change in one environment, or if an animal or plant finds itself in a new environment with selective pressures other than what it is used to, the animal may find that the gene combinations that previously served it well no longer do.

Genes are plastic; they can change. Such changes, called mutations, can affect how a particular trait will be expressed. Some mutations are deleterious; they can alter an organism in ways that are unfavorable or are actually debilitating or lethal. Some are neutral, but some, under the right conditions, can be beneficial.

An example of that sort of thing is a phenomenon called industrial melanism, where the most frequently observed coloration in two populations of moths changed, not once, but twice. Prior to the industrial revolution, peppered moths in both Britain and the United States were predominately a mottled gray in color. Black or melanistic individuals occurred occasionally. The moths are nocturnal, resting during the day on the trunks of trees. The moths are preyed upon by birds, which find them visually. When the tree trunks are encrusted with lichens, the grey moths are well camouflaged and are difficult for the birds to see. In contrast, the black moths stand out plainly. The birds acted as selective agents removing black moths, and black genes, from the populations. However, black genes either remained hidden within gray moths or reoccurred by frequent mutation; they never disappeared. With the coming of the industrial revolution, air quality deteriorated killing the lichens on the tree trunks. Against the bare black bark, the gray moths stood out and the black ones were more camouflaged. Consequently, the birds began preferentially selecting gray moths and the populations changed from predominately gray to predominately black. During the latter half of the twentieth century, when air quality standards had been enacted and air quality improved, lichens once again began growing on tree trunks, and the gray moths were once again at a selective advantage. The example of industrial melanism has been challenged as being apocryphal, but it clearly shows how natural selection works, and it is well described, along with similar examples, by University of Wisconsin biologist Sean Carroll (2007). In addition, journal articles have been written describing the phenomenon. And, it demonstrates natural selection rather neatly.

One note: individuals do not change. A peppered gray moth will not spontaneously turn black. But gray moths can carry genes which, when combined with the right genes in a moth with which it mates, can produce black offspring.

Darwin and others believed that evolution occurs in small but continuous steps bringing about a gradual change in a species of organisms. In time, a species would evolve into a different species. Furthermore, they believed that it was progressive and that it brought about continual improvement. This model has been challenged by Niles Eldridge at the American Museum of Natural History and the late Stephen Jay Gould of Harvard University, who have framed the hypothesis they call punctuated equilibrium. This model has it that once a species evolves, it remains stable and changes little until its extinction, which may be abrupt. Furthermore, if a species becomes fragmented in any way so that genes can no longer be exchanged among all of its members, the isolated fragment may quickly diverge from its ancestral species as mutations within it accumulate, especially if it is subject to different selective pressures than is the parent species. In other words, according to Eldridge and Gould, evolution is characterized by long periods of stability, or equilibria, which are punctuated by short, eruptive bursts of change. In this model, a species could remain unchanged as long as its environment remains stable but still give rise to descendent species. This model has been called punctuated equilibrium as opposed to the gradualism described by Darwin.

Both models may be correct. In the case of human evolution, gradualism seems to have occurred. Fossil remains of our ancestors generally show sequential change since our divergence from our common ancestor with the chimpanzee and bonobo, some four million years ago. Moreover, transitional forms, so-called missing links, when discovered, have generally been found to be intermediate between the earlier and later forms that gave rise to and descended from them, respectively.

Fossils were the initial clues to evolution. They appear to have become abundant around 550 million years ago. Older ones exist, but they are rare, most likely because organisms existing then did not have hard parts. Soft tissues are much less likely to be preserved. However, in rocks that formed then, fossils were suddenly abundant. Moreover, they were diverse, as if a variety of life-forms appeared suddenly out of nowhere. That period of time, known as the Cambrian period of the Paleozoic era, showed not only an abundance, but also a wide assortment of life-forms. Because of their sudden variety and diversity, the event has been called the Cambrian explosion. Furthermore, following the mass extinction at the end of the Cretaceous period and others—there were five such extinctions in total—there was a rapid diversification from the survivors. Such sudden developments of biodiversity are taken as evidence that evolution can occur relatively quickly. In addition, the suddenness of some die-offs, such as the Cretaceous meteor and the other great extinctions, offer testimony that disappearance can be rapid as well. The two, taken together, offer support for punctuated equilibrium.

Support for rapid changes in the history of life are often multifaceted. Using the comet impact at the end of the Cretaceous period as an example, evidence is not limited to the sudden extinction of the dinosaurs and their contemporaries. Additional substantiation is provided by a thin band of iridium-rich rock that has been dated as having been deposited 65 million years ago. Iridium is rare on the Earth's surface, but it is common in many celestial structures, including some that strike the earth. Moreover, a strike by an asteroid or other type of celestial debris is not an unusual event in the expanse of geologic time. Admittedly, large bodies striking planets is rare, particularly from the perspective of the human life span, but it does happen. Between 16 and 22 July, 1994, multiple objects, some of them pretty large, collectively designated as the Shoemaker–Levy Comet, collided with Jupiter. This is the first time such an event has been observed, but evidence on Mars and the Moon show that collisions occur periodically. Neither body has an atmosphere like Earth's, where space debris will burn up, nor does either have a living surface that will eventually heal the impact crater. Consequently, both are rich in craters where space debris has struck them.

Admittedly, the asteroid that ended the dinosaurs' reign was far too large to burn up in the atmosphere, and it did leave a crater. Otherwise, there would have been no way of knowing that it struck off the Yucatan peninsula. That crater has been discovered by remote sensing and gravitational measuring techniques that are outside of the understanding of most but the scientists who use them; we will not dwell on them here. But interested readers are encouraged to do their own research on the subject. And while it has been 65 million years since that impact, there have reportedly been a number of "near misses", one supposedly as recently as late as January, 2008, as celestial bodies have passed close to Earth, with close being a term relative to the distances encountered in space.

Despite the evidence, some scientists do not accept the idea that the dinosaurs' extinction was caused by an extra-terrestrial body. They argue that the close of the Cretaceous was a period of extreme volcanic activity. Since iridium is present in the molten middle layer or mantle of the Earth, volcanic activity could account for the band of iridium described earlier. In addition, some gases released by volcanoes can easily react with atmospheric water vapor to form acid rain, and others could lead to a problem we are experiencing today: global warming. It is also possible that the volcanism was triggered by the impact. Scientists do disagree with one another, as noted above. It is possible that we will probably never know for certain what caused the extinction of the dinosaurs, or future research may one day enlighten us. The fact remains, however, that scientists do not argue over the fact that something catastrophic happened. By the way, my own bias is toward the asteroid impact; I find that evidence more convincing.

Finally, the chart on page 8 represents the last 550 million years of biologic history. Students of historical geology and evolution have had to memorize the times and names of each of the geologic eras and periods, and they do give useful titles to which to refer. I include them for that reason only; I have no intention of trying to force them on anyone. But the dinosaurs' extinction is easily referred to as having happened at the end of the Cretaceous, and the Tertiery, the first period of the Cenozoic era that followed, is an easy reference period for the vast evolution, or radiation, of mammals once the dinosaurs were gone. Indeed, the Cenozoic is referred to as the Age of Mammals, while the Mesozoic was the age of Reptiles, although dinosaurs technically were not reptiles. They were more closely related to birds.

Geologic era	Period	Duration (MYA)[a]
Cenozoic	Quaternary	0–1.8
	Tertiary	65–1.8
Mesozoic	Cretaceous	144–65
	Juassic	206–144
	Triassic	248–206
Paleozoic	Permian	290–248
	Carboniferous	354–290
	Devonian	417–354
	Silurian	443–417
	Ordivician	490–443
	Cambrian	543–490

[a] Million years ago

The Paleozoic that preceded the Mesozoic was the age of invertebrates, animals without backbones, although fish were common enough during the latter part of it, amphibians were around at the end, and reptiles had originated by the very end. Collectively, the three eras and their component periods are known as the Phanerozoic Eon.

The massive die-off of 65 million years ago is often called the Cretaceous–Tertiary extinction, or the K–T extinction, K for the German°*Kreidzeit*" for Tertiary. Four others are thought to have occurred. Starting with the most recent they were: End of the Triassic, when 52% of marine genera died off, probably as a result of volcanism and consequent global warming; Permean–Triassic, when 84% of marine genera and 70% of terrestrial species were lost; late Devonian, when 57% of marine genera were lost; and the Ordovician–Silurian, which resulted in the extinction of 60% of marine genera. Each mass extinction was followed by a radiation as surviving species began spreading into previously unavailable territory and eventually niches.

Extinction is the eventual fate of all species of animals, and a major geological or astronomical event need not necessarily occur to cause it. For example, some species are restricted to a very narrow geological range, perhaps a single island. A change on that island, such as the arrival of a new competitor or predator, may trigger its extinction. Over the past 15,000 years, the most destructive predator and competitor has been ourselves. Current anthropological evidence has humans originating in Africa and spreading across the rest of the world. Each arrival of humans on a new continent generally has been marked by a major extinction.

A case in point is Australia, which, as is generally known, is the last holdout of the animals we identify as marsupials. Currently, there are no large marsupial carnivores. When humans arrived there, some 50,000–60,000 years ago, there were at least two: a marsupial lion that may have been built more like a bear than it was the so-called king of beasts, and the Tasmanian wolf. The latter was once extant over the entire continent of Australia. Either competition with or predation by humans, competition with dingoes, a true dog brought to Australia by humans, or some combination thereof drove it to extinction on the main continent. It survived on the island of Tasmania until the twentieth Century before succumbing entirely. In addition, Australia was once home to a large predacious lizard, similar to but larger than the Komodo dragon of southeast Asian islands. This, too, may have disappeared in consequence of human arrival.

North America also experienced a substantial extinction of large mammals coincidental with the arrival of humans. Humans arrived in North America most likely between 10,000 and 13,000 years ago, toward the end of the last ice age. Their arrival spelled the doom of the American horse, the giant sloth, the American mastodon and wooly mammoth, and others including possibly the cave bear, saber-toothed cat, and dire wolf. Some scientists have argued that those animals were unable to adapt to the end of the ice age and subsequent warming, but many disagree. The horse, saber-tooth cat, and dire wolf originally evolved in the Americas; they had probably experienced numerous climatic fluctuations over their existence. The giant sloth immigrated from tropical America. It was adapted to warm conditions. Indeed, the horse survived in Eurasia having arrived there by the land bridge that once connected Alaska with Siberia during the glacial period. The same land bridge brought humans to North America. Native North American animals, however, were not adapted to coexisting with humans.

Today, human-caused extinction is being compared to the five mass extinctions of the past. Indeed, the subdivision of the Quaternary period in which we now live, the Holocene epoch, might well be called the Homocene, the age of humans, our impact has been so great, and our activities have resulted in the extinctions of uncounted species of plants and animals. By over-hunting, destroying habitats, fouling the environment, and expanding our own numbers, we have eliminated many organisms, some of which may potentially have been useful to us. Moreover, we are now changing the planet we live on in ways that could possibly end up causing problems for ourselves.

Extinction is the final step in evolution, just as death is the final step in life. In the subsequent chapters, we shall examine how life has changed on earth and how it has affected and been affected by us.

Reference

Carroll S (2007) The making of the fittest: DNA and the ultimate forensic record of evolution. W. W. Norton & Company, New York

Chapter 2
What's God Got to Do With It?

In 1633, the Tuscan scientist Galileo Galilei was brought before the Roman Inquisition to be tried for heresy. Galileo's crime was publishing an account of heliocentricism, the concept that the Earth is neither the center of the universe nor is it fixed in place, both of which are contrary to scripture. He was pronounced guilty and condemned to house arrest for the remainder of his life.

Today, not only is the fact that the Earth is neither fixed in place nor the center of the universe not only known to be the case, it is accepted by virtually every major institution on the planet, including the church that tried Galileo for heresy nearly four centuries ago. That church, by the way, continues to survive even though scripture turned out to be technically incorrect, although it is admittedly facing different challenges today.

To pre-industrial Europeans, the idea that the "world is firmly established, it cannot be moved," (1 Chronicles 16:30) was perfectly reasonable. From their perspective, the Earth never changed in position, while one could easily see the sun and moon circling it. Any other interpretation was counter-intuitive. Ironically, though, the idea of heliocentricism had occurred in India, Greece, and the Middle East well before the origin of Christianity, and the Polish scientist Nicolaus Copernicus had demonstrated it nearly a century before Galileo was condemned. Moreover, with no idea of the cosmos nor the geophysical forces that shaped the planet, it was also perfectly reasonable for pre-industrial people to believe that the world and everything on and around it were constructed by super-natural beings, and physical phenomena those people could not understand were appropriately taken to be signs of those beings' existence. A case in point would be weather. Lightening, for example, has long been taken to be a sign of God showing his displeasure, and even as recently as 2005, some people claimed that Hurricane Katrina flattened New Orleans because of God's anger at the sin that was said to be rampant within the city. We know, of course, that lightening is nothing more than an electrostatic discharge that occurs when a charge builds up in one part of the atmosphere and is released to another. In the case of cloud to ground lightening,

B. Marcus, *Evolution That Anyone Can Understand*,
SpringerBriefs in Evolutionary Biology, DOI: 10.1007/978-1-4419-6126-6_2,

the discharge often finds the shortest possible distance between its origin and destination, usually a low point in a cloud and a tall structure like a building or tower on the ground. In natural areas, suburban developments, and the past, the tallest structure could very well be a tree, and lightening striking a tree can knock it over or set it afire. If houses or other lower structures are nearby, a struck tree could fall on or spread fire to them, lending credence to the idea that the resident is being somehow punished.

The late Stephen Jay Gould of Harvard University showed the fallacy in that reasoning by pointing out that in medieval Europe, the tallest structure in any village was usually the village cathedral, and lightening usually hit it more often than any other building, including the village tavern or brothel. By the same token, hurricanes and other weather phenomena follow random, though often cyclic, patterns. Hurricane Katrina was no more indicative of the wrath of God than is a tornado or lightening striking a church; it was a chance event, just as any destructive storm or other natural disaster is. Indeed, given New Orleans' location and topography, it is surprising that it has not been the target of many more such natural disasters. In the end, however, the carnage that resulted and the innocent people who suffered along with the guilty were simply results of misfortune, a city that was largely built below sea level, and an infrastructure that was not up to the task demanded of it. It was not collateral damage from the rage of a vengeful deity.

As science has progressed and knowledge has grown, understanding of the movement of the planets and weather, among other natural phenomena, has improved. Today virtually nobody would insist that the Earth is the center of the universe, and most people understand that there are natural laws that govern weather; it is not an emanation of some angry or capricious divinity. The same holds true for evolution. As I will try to show in subsequent chapters, the growth of biological knowledge has shown evolution to be a natural phenomenon, and there are observable and verifiable realities that affect it. Nevertheless, evolution seems to be the cause around which religious dogmatists have rallied. Their logic appears to be that if evolution is somehow disproved, their argument wins by default, a case of the non sequitur not A, therefore B. In contrast, equally dogmatic atheists argue, often just as passionately, that the existence of evolution is somehow counter to the existence of God. Ironically, both arguments are incorrect. Religion and science are not mutually exclusive entities; they are unrelated subjects that serve totally separate human needs. The veracity or inaccuracy of one has nothing to do with the other.

The religious argument against evolution often includes the assertion that the bible provides a thorough, literal account of and explanation for the origin of life and the current state of biodiversity that exists on the planet. Moreover, those who advocate that contention also maintain that the bible is an accurate description of history and an unambiguous account of God's revelation. A careful reading of the bible, however, shows that it is anything but unambiguous. Case in point, Chapter 3, Verse 16 of the Gospel of St. John in the King James edition of the Holy Bible states:

> "For God so loved the world, that he gave his *only begotten Son* (my italics), that whosoever believeth in him should not perish, but have everlasting life."

In the same bible, Chapter 6, Verse 2 of the book of Genesis states:

> "That the *sons of God* (my italics) saw the daughters of men that they *were* (sic) fair; and they took them wives of all which they chose."

One passage states clearly that God had only one "begotten Son," while the other states equally clearly that God had more than one son. If one is to interpret the bible literally, both passages cannot be correct. By the same token, two different accounts are given of the genealogy of Jesus in Matthew 1:2–16 and Luke 3:23–38. Neither of these can be correct, however, because both have Jesus as a descendant of King David through Joseph, presumably his father, while it is a major tenet of Christianity, and Islam as well, for that matter, that Jesus was born of the Virgin Mary. The same gospels that list his genealogy state that he was conceived by the Holy Spirit; no mortal father was involved.

These objections to the bible disappear if one allows for an allegorical interpretation. However, those who deny evolution on religious grounds usually demand to have the bible interpreted literally. With that being the case, the description of biological creation is also contradictory. The best-known account is given in Genesis 1:11–27 that has God creating first life on the third day. The order of creation is grass, herb yielding seed, and fruit trees. On the fifth day god created fowl that flies, great whales, winged fowl (again), cattle, creeping things, and beasts of the earth, and on the sixth day he created man, *male and female* (my italics). But there is another account of creation that is divided among Genesis 2:7, 19, 21, and 22. It has God creating man, or Adam, first, beasts of the field and fowl of the air next, and finally woman, Eve. Neither version describes, for example, the creation of arboreal mammals, beasts of the trees as it were, or bats, beasts of the air. Likewise, animals like ostriches and emus, fowl of the earth, penguins, fowl of the ice, and many other forms of life are not mentioned. Numerous other examples of biblical inconsistencies are described by journalist Christopher Hitchens' (some would call it blasphemous) book *God is not Great* (Hitchens 2009). Again, none of these apparant contradictions present a problem if one allows for an allegorical or otherwise non-literal interpretation of the bible, but in a literal interpretation, they present a conundrum.

Parenthetically, I have often found that dogmatic creationists will deny that any contradiction occurs in the bible. They will argue, for example, that the sons of God in Genesis are really angles, or that the two stories of creation are really the same story. This in itself is contradictory because it renders one of the accounts at best metaphorical. Calling metaphor literal truth is disingenuous.

The specificity of creation as presented in Genesis has led some to identify a precise time of creation and, consequently, the age of the earth. Perhaps best known of these was James Ussher (1581–1656), an Anglican bishop in Northern Ireland, who calculated the origin of the earth to have occurred just about six thousand years ago. Owing to a gap in the biblical record between the end of the Old Testament and the beginning of the New Testament, Ussher was forced to rely on Roman and Greek history as well as the chronology given in the bible. In addition, his calculations were no doubt made more difficult by the conversion from the Julian calendar to the Gregorian calendar, which occurred in 1582.

Ussher was by no means the only person to have attempted to date the beginning of the earth, and his calculations are pretty consistent with those of other scholars. For example, the Jewish calendar holds that the age of the earth is a bit less than 5800 years; it corresponds with the earth having been created in around 3760 BC. According to the Chinese calendar, the age of the earth is a bit over 4,700 years.

Not all creationists take Ussher's calculation as precise. Some acknowledge that the earth may be as much as 10,000 years old, but none of these accepts the antiquity of the planet that science has shown to be the case. A young earth, however, introduces many complications given the physical realities of the planet. Some of these are considered in a later chapter.

The discovery and elaboration of DNA is further evidence against any kind of special creation. If, for example, one considers the DNA of great apes, two of them, chimpanzees and bonobos, are biochemically more like humans than they are like other apes, which means that they are more closely related to humans than they are to other apes. This is counter-intuitive; superficially, structurally, and developmentally, all apes appear to be more like one another than any of them is like us. Still, their DNA says otherwise. In addition, among ourselves, there is so little variation in DNA that the concept of race, long an issue in our particular society, turns out to be, for all practical purposes, nonexistent. In spite of this, creationists have argued that DNA is itself a product of creation that it and its functioning are so complicated that in no way could it have resulted from random evolution; its development had to be directed. They have called this concept intelligent design. That, in reality, is little more than a renaming of creationism or what has been euphemistically called creation science. Parenthetically, the same argument about complexity was used more than a century ago. Back then, however, it stated that the diversity of life was so great that it could not possibly have resulted from random evolution; it must have been the result of deliberate creation.

Throughout nature, there are many cases where the design would appear to be somewhat less than intelligent. An example of this might be sexual cannibalism, as described by biologist Marty Crump (2007). Dr. Crump talks about, among other things, female spiders that eat their mates immediately after mating. Indeed, the males often appear to offer themselves as meals. From a biological perspective, there is arguably some sense to this. A well-nourished female is more likely to produce viable eggs. By sacrificing himself to his mate, a male may insure that by providing nourishment, the genes in the sperm he just donated will stand a better chance of making it into the next generation and his DNA will be perpetuated. Perhaps the same logic applies to some species of preying mantises, where the female also dines on her mate. But more bizarre among these insects is that sometimes the female eats the male during copulation, and sometimes she will bite his head off prior to mating. Regardless, the decapitated male may still succeed in fertilizing his murderess. Indeed, Crump states that decapitation actually facilitates sperm release, in which case, once again, the male's genes stand a better chance of surviving into at least the next generation. There are, admittedly, other possible explanations for this behavior, but it happens frequently enough, and the process sounds pretty grisly, especially as described by entomologist L.O. Howard in 1886 (Gould 1984).

If sexual cannibalism is weird, how would one describe parental cannibalism? Some species of spiders, pseudoscorpions, and earwigs practice matriphagy, where young eat their mother. In the words of Gould, greater love hath no woman. The practice sometimes occurs when food is scarce, and it is believed that it prevents cannibalism by the young among the spiders and pseudoscorpions and greater survival among the young earwigs. In both cases, the mother's sacrifice is an attempt to ensure the survival of her offspring, and in the grand scheme of nature, such that it is, is it any different from a doe's giving herself up to a pack of wolves so that her fawn might survive? Still, one would think that a benevolent deity would have found a kinder way of solving the problem.

Perhaps less bizarre but still puzzling eating behavior occurs in two species of colobus monkeys. The black and white colobus monkey and the red colobus monkey are inhabitants of Africa, where they both live in the forest canopy and survive by eating leaves. Oddly, the leaves they consume are full of toxic chemicals that other monkeys cannot tolerate. Colobus monkeys, however, have a complex digestive system and bacteria living in their stomachs that allow them to decompose the intricately structured chemicals and toxins in the leaves they eat. However, they do not appear to do it with impunity. Colobus monkeys often sit around lethargically for hours after they eat, and they have been described as looking dyspeptic. Some colonies of red colobus monkeys in Zanzibar living in or near what have been described as perennial gardens near human habitations, however, have learned that by eating charcoal, they can relieve the dyspeptic symptoms. Moreover, these colonies live in greater population densities than one finds among forest dwelling groups, suggesting better survival (Strhsaker et al. 2004).

Charcoal is not a normal part of an animal's diet. In fact, only red colobus monkeys and humans eat it, and those monkeys that do live in the proximity of humans. Why they began, nobody knows, but in doing so, they have apparently improved their lives. Eating toxic leaves does not make much sense, unless one considers that the monkeys doing it do not have to compete with other leaf-eating animals. It can be argued that God gave the colobus monkeys the digestive equipment to deal with such a toxic diet, but if he did, why did he not equip them to live more comfortably with the consequences of it. Moreover, why did he give only a few of them, which coincidentally live near humans, the means to relieve those consequences?

An even more strange behavior pattern is coprophagia, the eating of feces. The natural decomposition of these materials by bacteria and fungi is, of course, a necessary process in returning essential plant nutrients to the soil. That is one reason why rotted animal manures are used extensively as fertilizers and soil builders in gardening and organic farming. Furthermore, the consumption of fecal matter by insects like dung beetles and fly larvae is understandable enough because it facilitates the decomposition process, although one might take pause upon seeing this kind of behavior in brightly colored, attractive butterflies. However, coprophagia by mammals is, in contrast, considered to be aberrant if not completely disgusting. Yet it happens commonly. Rabbits, for example, typically eat their feces, often plucking the material directly from the anus. Dogs routinely eat the feces of other animals, and a couple of our biological near relatives, orangutans and gorillas, eat their own and those of conspecifics. In the latter two, this was first observed among confined animals in zoos, where it was assumed to be a response to crowding or some kind of stress. However, it has since been observed in the wild where stress did not appear to be a factor.

From a biological standpoint, the behavior is not at all odd. Dogs are scavengers, and fecal material is known to contain nutrients. It is often an easy meal for them. Rabbits, orangutans, and gorillas have diets that contain much coarse and difficult to digest vegetable material. A single passage through their digestive systems is often not adequate to break food down completely; a second passage provides greater access to many nutrients, including some vitamins that are manufactured in the large intestine. But the consumption of another animal's feces also introduces the possibility of encountering intestinal parasites or other infections. Worms are commonly found in animal intestines, any many shed their eggs with their hosts' feces, counting on another animal to eat them. It would seem to be a far more clever idea on the part of a designer to have either equipped animals that eat coarse diets with a digestive system that could handle such material, or else provided them with a desire for and access to more digestible fare.

Such behaviors are usually not part of our own nature; consequently, they have been historically considered abnormal in other animals as well. Yet they are common. Even homosexuality, considered an abomination by many of us, is routine enough in nature. Indeed, the animal that practices it most routinely is the bonobo, which, as mentioned previously, is, along with the chimpanzee, our closest animal relative.

One sees other apparent natural paradoxes among mammals, ourselves included. During embryonic development, for example, we produce and then abandon gill slits and other primitive features. Our ear bones initially form in our lower jaws and then migrate, rather than forming in our middle ears, where they are eventually used. Our skeletons originate as cartilage models, like the skeleton of sharks. That by itself would not appear to be really odd, unless one realizes that the part of the mammalian skeleton where this does not occur, that is where bone is produced without replacing cartilage, the upper skull, represents a part of the skeleton that sharks lack. Much of this seems energetically wasteful; pregnancy could be shorter and less energy intensive if some of these seemingly unnecessary steps did not occur. Additionally, in humans specifically, the upright posture that served us so well during the hunting-gathering days of our ancestors now becomes problematic as we age. The stress on our knees and hips makes them targets for arthritis, and lower back pain and crippling resulting from herniated intervertebral disks is often the result of stress on our spines.

Returning to reproduction for the moment, it can conceivably be argued that sexual cannibalism is actually part of design. After all, it provides the female with nourishment that she can use, but if that is the case, one would have to ask why it is so rare. Moreover, why are there ways around it? Among some orthopterans (grasshoppers, crickets, etc.) male glandular secretions are often presented to the female to be eaten. Among some predacious flies, the male offers captured prey to the female for her consumption. Both of these behaviors satisfy the female's need for nourishment without sacrificing the male in the process, thus freeing him, potentially, to mate again and increase his gene frequency in subsequent generations. Consequently, while sexual cannibalism apparently works toward insuring gene survival, it is less than a perfect method of maximizing gene presence in the succeeding generation and a less elegant design than that of the orthopterans and flies. Moreover, if all three are the work of a single designer, why are there three separate designs? In the case of matriphagy, once again, if it is a good design, why do we not see more of it? Besides that, it, along with sexual cannibalism, seems rather sadistic.

The plant and animal kingdoms are replete with bizarre strategies for reproduction and survival that defy rational explanation other than that they work. If they are the result of design, one could credit the designer with something of an odd sense of humor. Representative of that is what has been called the "prostitute orchid" of the tropics. In this plant, the reproductive structures of the flowers are arranged into a mimic of a female wasp. They even produce a chemical that duplicates the sex-attractant pheromone of that wasp. Males of the wasp species mount the decoy in the flower, attempt to copulate with it, and end up getting nothing for their efforts other than being dusted with the orchid's pollen. Whatever sperm the wasp deposits and energy he expends are wasted; there is no perpetuation of his genes, not even a meal of nectar that most pollinating insects get from the flowers they visit. Worse, while covered with pollen, he can be lured to another orchid blossom on which he again wastes his efforts, but the orchid gets pollinated in the process. One can argue that from the orchid's perspective, if this is a case of design it is indeed intelligent, but from the wasp's, it is rather mean.

A practice sometimes used by logicians in trying to decide which of multiple explanations is most likely correct is known as "Occam's razor." Ascribed to 14th century Franciscan friar William of Ockham, it is another way of saying, "When you hear hoofbeats, think horses, not zebras," and can be boiled down to "the simplest explanation is usually the best one." Three separate, deliberate designs for feeding female invertebrates, one of which is described at best as bizarre, seems a bit complicated in contrast to each of these strategies having arisen as a result of random mutation and subsequent selection. Moreover, the mass of arthropods eating their mother and the frustration of the male wasp hardly seem to be reasonable ends of a creator generally held to be benevolent. And while some in the religious community insist that such natural oddities could not have occurred in the absence of a creator, others disagree. Indeed, the General Assembly of the Presbyterian Church, the Central Conference of American Rabbis, and more than 10,000 Christian clergy representing multiple denominations have drafted letters supporting the compatibility of evolution theory with religious teachings. Furthermore, on October 22, 1996, Pope John Paul II reaffirmed the Vatican's earlier position that evolution and faith are not irreconcilable. Consequently, anyone who objects to evolution on the basis of faith has to recognize that he or she is not subscribing to a unanimous agreement. Indeed, it might well be the minority position.

The contrary argument that the reality of evolution somehow disproves the existence of God or is contrary to faith is also something of a *nonsequitur*. While some may argue correctly that there is no tangible evidence to support the existence of God, the counter argument, the absence of evidence is not evidence of absence, is equally true.

Indeed, science can neither prove nor disprove the existence of God. Its realm of study is limited to the natural world where tangible evidence can be gathered and hypotheses tested. God, by definition, is supernatural. His existence can be accepted or rejected only as a matter of faith, and faith is the acceptance, or rejection, of something in the absence of evidence. Visually, it would look something like this.

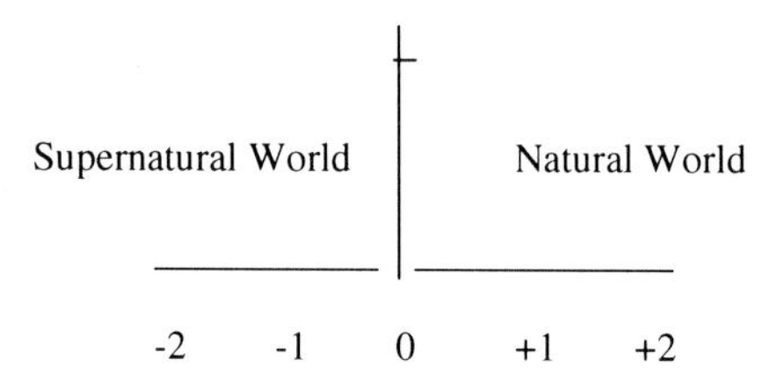

The above axis is a hypothetical representation of our world. The region to the right of the zero line represents the concrete, hands on world in which we live. This is the realm of science, where objects can be explored, measured, and manipulated. The left of the zero line represents the world of religion. Science cannot enter it, and so it can never be described in anything but the abstract. Moreover, our sense organs, which evolved in the natural world to perceive phenomena within the natural world, are incapable of observing events anywhere but within the natural world.

Syndicated columnist Cal Thomas has argued that the inability of science to prove God's existence demonstrates the weakness of science. Perhaps so, but could not a nonbeliever similarly argue that the inability of the religious to accept scientific fact on the argument that it disagrees with their beliefs represents a weakness of faith?

Just as there are people of faith who accept evolution, there are people of science who, coincidental with their acceptance of evolution, have faith. Cases in point would include Brown University biologist Kenneth Miller, co-author of a best-selling high school biology textbook, who points out that "... religious scientists look for God in what science does understand and has explained," rather than describe everything that has not yet been explained by science, along with much that has, as divinely created. Along the same lines, Francis Collins, director of the Human Genome Project and now director of the National Institutes of Health, has stated, "God's domain is in the spiritual world, a realm... must be examined with the heart, the mind, and the soul." University of Cambridge, England, paleobiologist Simon Conway Morris has written books about the harmony between faith and evolution. Not all scientists are believers, of course. Some are vocal atheists, but that is their position as individuals. There are also atheist economists, musicians, and accountants. Science itself does not take a stand on religion, other than on points where religion chooses to contend.

Spirituality appears to be a universal human characteristic. Undoubtedly there have been doubters in every society in all ages, but in general, belief in a supernatural being or beings appears to be characteristic of practically every civilized, barbaric, and savage culture in the history of mankind. Ancient Rome and the recent Soviet Union are among the few in which atheism was openly embraced. Indeed, there have been recent books, such as Dean Hamer's *The God Gene: How Faith Is Hardwired into Our Genes*, that suggest that spirituality is part of our genetic heritage and is necessary for our social survival, an indeed ironic piece of evolution if correct (Hamer 2005). It might suggest that we created God and not the opposite, sort of supporting Voltaire's "If God did not exist, it would be necessary to invent him." But at the risk of being repetitive, the existence of God and the veracity of science are not mutually exclusive concepts. Furthermore, the actuality of evolution does not invalidate the existence of God, and disproving evolution would not confirm the existence of God. The natural and supernatural domains, if the latter does indeed exist, are separate and will probably remain so for at least the foreseeable future.

References

Crump M (2007) Headless males make great lovers: and other unusual natural histories. University of Chicago Press, Chicago

Gould SJ (1984) Only his wings remained. Nat Hist 93:10–18

Hamer D (2005) The God gene: how faith is hardwired into our genes. Anchor Publishing Co, Harpswell

Hitchens C (2009) God is not great: how religion poisons everything. Twelve, New York

Strhsaker TT, Cooney DO, Siex KS (2004) Charcoal consumption by Zanzibar red colobus monkeys: its function and its ecological and demographic consequences. Int J Primatol 18:61–72

Chapter 3
But It is Only a Theory

On 16 July 1945, a team of scientists assembled near Alamogordo, New Mexico to watch the first test of the atomic bomb. Given the research that had gone into the atomic bomb's production, most of the scientists probably expected the test to go off as planned, but nobody really knew for certain. Some worried that absolutely nothing would happen, others feared that an earth-destroying series of cataclysmic explosions would be touched off. Till that time, nobody had ever seen an atom. Its existence was entirely theoretical. That must have been of enormous comfort to the citizens of Hiroshima and Nagasaki, Japan, upon whom, on 6 and 9 August 1945, respectively, atom bombs were dropped. Those were the only times atomic energy was used in warfare, and debate over whether or not it was justifiable continues to this day. Since then the atom has been put to more constructive use. It is commonly used today to generate electricity and to drive battleships. Nobody doubts the existence of the atom, and scientists work with them and with subatomic particles all the time. But to date, nobody has seen an atom, and scientists still speak routinely about the atomic theory.

The word theory is by no means precisely defined. In informal conversation, it is often used to describe any idea, whether based on fact or otherwise. It is often thought to be part of a hierarchy that begins with a hypothesis, often defined as more or less an educated guess, and then continues through theory and finally to law. The popular concept seems to be that theories are somehow questionable or flawed. In science, a theory is based on verifiable fact and is often an explanation of that verifiable fact. In evolution, it is a verifiable fact that fossils occur in sedimentary rocks and that fossils vary in appearance through time. Evolution theory is employed to explain that phenomenon.

Theories are not immutable. The atomic theory, for example, has changed substantially since it was initially proposed. The modern atomic theory had its origin perhaps nearly two centuries ago when the English scientist John Dalton found that matter appeared to be made up of tiny particles. He suggested that those particles were elementary and in no way divisible. Toward the end of the

B. Marcus, *Evolution That Anyone Can Understand*,

SpringerBriefs in Evolutionary Biology, DOI: 10.1007/978-1-4419-6126-6_3,

nineteenth century, experiments had shown that the atom was indeed divisible into so-called subatomic particles. The arrangement of those particles was inferred to mimic a solar system with the smallest particles, the electrons, orbiting a central nucleus. This model has been modified over the past century as evidence accumulated that has shown that additional subatomic particles exist, and that the planetary arrangement did not account for all observable atomic properties. However, the atom continues to be theoretical even though the existence of atoms themselves is factual. Even if subsequent research shows our current model of it to be incorrect, atoms will continue to exist. Debunking the current theory will not change that. Yet nobody seems to be demanding that an alternative to atomic theory be given equal time in chemistry classes.

Theories are important in many aspects of scientific advancement. In medicine, for example, a particular illustration would be the germ theory, which has it that illness is caused by living entities that cannot be seen. In one form or another, this idea has been around for more than two millennia. It began to take on its more modern form as a result of the work of the English physician Edward Jenner who, in 1796, was the first person to successfully use vaccination for the prevention of smallpox. In actuality, vaccination had been attempted previously. Physicians at the time knew that anyone who survived a smallpox infection appeared to be protected from the disease for the rest of his or her life. Some attempted to introduce the material from the pustules of smallpox victims who had mild infections into healthy individuals. The practice usually resulted in the recipient becoming ill, sometimes surviving and sometimes not. Jenner, however, noticed that the dairy farmers and farm workers who handled dairy cattle in his part of England often developed a milder infection that rarely, if ever, killed, but still protected the victim against developing smallpox. Eventually known as cowpox, this infection was usually limited to the hands of the worker. Jenner reasoned that exposure to and infection by the cowpox agent protected people from smallpox. He removed pustule material from the hand of an infected young milkmaid patient of his and introduced it, by skin cuts, into the son of a local farmer. The boy became mildly ill, but he recovered. A month and a half later, Jenner inoculated the boy with smallpox by the same method. One can imagine that he held his breath waiting to see what would happen. If his hypothesis was incorrect, the boy would become ill and probably die, and Jenner would be considered a murderer. As history now tells us, however, Jenner's reasoning was sound; the boy never developed smallpox; the experiment was a success.

We know today that smallpox is caused by a virus. Cowpox is also caused by a virus, one that is similar enough to the smallpox virus to provoke an immune reaction that is protective against smallpox. Jenner had no knowledge of viruses. Although the microscope had been invented perhaps 120 years earlier and it is probable that Jenner was acquainted with the existence of microorganisms, he could not have known about the existence of viruses. They are simply too small to have been seen with the crude instruments of that time, even though it is most likely that virus-containing material had been examined. Still, by observation and reasoning, Jenner was able to, we presume, infer that some kind of infective particles were responsible for smallpox and cowpox, and that exposure to the latter conferred immunity against the former.

The same kind of reasoning was probably employed by Ignaz Semmelweis, a Hungarian physician who also contributed to the germ theory. There are a number of versions of exactly how Semmelweis had his epiphany, but they all vary around a common theme: the rate of post-delivery infection and death among women in maternity clinics. In particular, Semmelweis observed that the highest rate appeared to be in clinics staffed by students who, prior to attending the deliveries, had participated in dissections, presumably of cadavers. The students had washed their hands with soap and water, but Semmelweis initiated the use an antiseptic solution, after which post-delivery death rates in their clinics plummeted.

In spite of the apparent cause and effect relationship between hand sanitation and infection prevention, not all physicians, including the one that headed the clinic where Semmelweis was employed, were quick to accept his conclusion. Such too is the nature of scientific theories. The identification of bacteria as the causative agent of infection had still not been demonstrated, and in general, scientists are conservative thinkers. They are skeptical by nature. Thus, before a theory becomes universally accepted, it has to be supported by abundant evidence. However, between the work of Jenner and Semmelweis, evidence was beginning to accumulate but not yet enough to be entirely convincing. During the 1860s, two more science pioneers, biologist and chemist Louis Pasteur in France and physician Joseph Lister in Scotland, added substantially to the pile.

Pasteur is best known for the process that bears his name: Pasteurization of milk. It involves heating milk to kill microorganisms that, among other things, hasten milk spoilage, but more importantly cause tuberculosis. However, like any good Frenchman, Pasteur's concern was with wine. Pasteur had observed that as wine, and beer for that matter, spoil, they become contaminated with microorganisms, bacteria specifically, other than the ones that are desirable for appropriate fermentation. By Pasteur's time, microscopes had reached a sufficient point of sophistication that allowed scientists to see many bacteria. Pasteur demonstrated that by heating the wine to a critical point, the contaminating bacteria could be eliminated, and the wine could be preserved. The process has since been extended to not only milk, but to other consumable products as well.

Perhaps more important in the evolution of the germ theory was the connection Pasteur made between bacteria and disease. In the early 1860s, the French silk industry was in crisis. Silk worms were falling ill and dying before maturing to the point where they spun their cocoons. Pasteur discovered bacteria in the sick worms that were absent in healthy ones. When the healthy worms were exposed to the bacteria, they became ill. More or less simultaneously, French cattle farmers were having trouble with anthrax. Pasteur discovered that there were differences in the severity of the disease, and that cattle that survived the milder forms became immune to the disease as a result, much like survivors of cowpox were immune to smallpox. He developed a vaccine using a non-virulent form to immunize cattle against the disease. Even more impressive was his development of a vaccine for rabies. Rabies is caused by a virus, which at that point was still unable to be seen under a microscope. Even so, Pasteur was able to generate symptoms of rabies in healthy animals by inoculating them with body fluids from sick ones. From there he was able to develop an attenuated version of the virus, which he tested successfully on a human in 1885. Pasteur had also demonstrated that microorganisms are present in and can be passed through the air.

Pasteur accomplished more in his lifetime than most of us would accomplish in a dozen or more. His work would probably have been sufficient confirmation of the germ theory, but still more evidence was added by Lister and the German microbiologist Robert Koch.

As mentioned earlier, Lister was a Scottish physician. A surgeon actually, Lister took an interest in post-operative wound infection and the mortality that resulted from it. He knew of Pasteur's work, particularly that relating microorganisms to fermentation, and he reasoned that wound sepsis could result from bacteria contaminating a wound. He was also aware that carbolic acid, today known as phenol, had been used to treat sewage and, as a somewhat serendipitous result, had helped reduce a certain parasite of cattle. He used carbolic acid to disinfect wounds with success. Post-operative infection became less and less of a problem at his Glasgow Royal Infirmary, and his technique was readily adopted, particularly in Germany. However, British and American surgeons, while satisfied with the technique, remained indifferent, perhaps even antipathetic to the idea that microorganisms were the cause of sepsis.

The final nail in the germ theory was probably driven by Robert Koch. Originally trained as a physician, he became interested in the role of microorganisms in disease and developed techniques that are still used today for culturing and staining them. He worked specifically with bacteria, particularly those that caused anthrax and tuberculosis, but he discovered other disease germs as well. Koch defined four steps, known today as Koch's postulates, which definitively relate a particular microbe to a specific disease. They are as follows:

1. The microbe responsible for a disease must be found in every case of the disease.
2. The microbe must be isolated and grown in pure culture, i.e. no other microbe can be present in the culture.
3. When the microbe in pure culture is introduced into a healthy but susceptible host, the disease must be reproduced.
4. The microbe must then be recovered from the experimentally infected host.

Koch's contribution to medicine paved the way for subsequent discoveries, such as penicillin and other antibiotics. His work not only cemented the germ theory but also paved the way for the level of medical treatment for infection that we enjoy today, but it was not perfect. It did not work for viral diseases.

Viruses cause many diseases, from the common cold to influenza, polio, measles, mumps, and chicken pox, not to mention the smallpox that was so problematic before Jenner. As mentioned previously, a virus also causes rabies. Viruses are not only too small, usually, to be seen with standard light microscopes, but they are also very difficult to culture. Viruses are described as obligate intracellular parasites, which means that they can grow only inside a living cell. Unlike most bacteria, they cannot be grown on culture medium; they must be grown in tissue cultures. In Koch's time, culturing human or other kind of living tissue was not possible. Consequently, those who blamed disease on bad air, devils, or impious thoughts and deeds still had viral diseases to support their claims.

In the 1890s, the Russian botanist Dmitri Iwanowski demonstrated the existence of viruses. Working with diseased tobacco plants, he showed that a homogenate of the plant leaves when passed through a filter known to be able to hold back the smallest known bacteria was able to cause the disease in healthy plants. Subsequent work by other investigators demonstrated that other diseases could be caused the same way. It was probably not until the invention of the electron microscope in the 1930s that viruses were first seen, but evidence still supported their existence. When AIDS was first observed, there were, again, those who insisted that it was not a normal disease but God punishing people whose behavior offended him, specifically homosexuals and intravenous drug users. When innocent heterosexuals and non-drug users started developing the disease, it became obvious that an infectious agent was the cause, and researchers armed with knowledge of the germ theory eventually isolated the HIV virus. Even additional infectious particles, viroids and prions for example, are being studied today, knowledge of their existence probably coming as a result of the recognition that the diseases they cause are infectious.

Of course, there are many diseases that are not a result of infection. Poor nutrition, heredity, gene malfunction, and improper immune system activity are responsible for many forms of human suffering, but none of these refutes the many illnesses that do result from infection. In short, the germ theory works, despite its not being a uniform explanation for all maladies. Should our referring to it as only a theory necessitate alternative explanations for germ-caused disease?

Yet another scientific theory is the cell theory. Normally credited to the work of three nineteenth century Germans, Theodor Schwann, Matthias Schleiden, and Rudolph Virchow, it had many more contributors. The pioneering work of Robert Hooke in England and Zacharias Jansen and Anton von Leeuenhoek in Holland paved the way for many of the later investigators. The cell theory basically says that all living organisms are composed of cells, and that all cells come from pre-existing cells. At the time the theory was framed that had been found to have been the case, and we have seen no exceptions to it since. Thus it stands to reason that if any form of life is currently unknown to us but is discovered in the future, it will be found to be composed of cells. That prediction has held in the past; there is no reason to believe that it will not continue to be the case. However, like all theories, it is potentially falsifiable. Conceivably, there is some non-cellular or acellular form of life hiding out somewhere on this planet, perhaps in a deep ocean vent, waiting to be discovered, or perhaps we will one day succeed in traveling to another planet on which life exists, where we will find something that does not conform to our cell theory. Not that I would recommend anyone's holding his or her breath waiting for either scenario to play itself out, but they remain possibilities. If one or both does eventually prove to be a reality, then the cell theory will have to be modified to accommodate the new evidence. All of the forms of life so far encountered have been cellular. That will not change. But potential falsifiability is a reality, and a good theory must be sufficiently plastic to accommodate it when it occurs. If it occurs too often, then the theory may have to be abandoned, but I do not see that happening to any of the theories discussed in this chapter nor to evolution.

Like the theories discussed in this chapter, evolution is supported by verifiable fact, and it has been used to make predictions that have turned out to be correct, another characteristic of a good theory. A case in point is the Madagascaran orchid *Angraecum sesquipedale*, which has a nectary over 11 inches long. Darwin predicted that a moth with a proboscis long enough to probe the orchid's nectar and to pollinate the orchid existed. He never saw it, but 50 years or so later, it was discovered.

Evolution also explains much in biology that would otherwise appear to be total chaos. Writing in *The American Biology Teacher* 2 years before his death in 1975, geneticist Theodosius Dobzhansky, most recently of Rockefeller University, stated, "Nothing in biology makes sense except in the light of evolution". Given all that has been learned in the past 30 years, the statement is even more correct now than it was when Dobzhansky wrote it. The Human Genome Project was not even a germ of a thought then. Now that it has been completed and many other means of studying biochemical genetics are available, the strength of evolution theory is more solid than it has ever been. Moreover, discoveries in Africa have even elucidated the history of our own species, something that has always been and no doubt will always continue to be, a bone of contention between those who accept evolution and those who do not. But missing links have been found, and our lineage, while admittedly not known perfectly, is clearer than it ever has been. Undoubtedly, more evidence will accumulate as more studies are conducted. Some may support the theory as it now stands, and some may force reconsideration. New evidence is always welcome. The more we have the closer we can be to getting it completely right. But evolution will always remain a theory, a proven one but a theory nonetheless. In science, that is not a weakness.

Chapter 4
Icefish and Other Genetic Anomalies and an Argument for Vestigiality

If one watches the embryonic development of a mammal, say a mouse, one sees a gradual increase in complexity as a single cell multiplies and develops into tissues and organs. But it does so in steps. The heart, for example, rather than forming as an intact entity begins as a blood vessel that folds to form two distinct chambers, which are characteristic of the heart of a fish. One of the chambers then divides into two so that three are present, which is much like the heart of reptiles. Finally, the undivided chamber is partitioned so that the four chambers characteristic of a mammalian heart finally form. The process appears to be one of adding complexity to simpler, previously developed structures. In fish, blood vessels run from the heart to the gills where carbon dioxide is exchanged for oxygen. Those blood vessels exist in mammals early in development, but they become modified to carry blood from the heart to the head, arms, lungs, and trunk as development proceeds. From the construction point of view, this is not the most efficient way of doing things; it is analogous to building a diesel locomotive by first building a steam engine and then re-engineering it into a diesel by using the original materials and reworking them. Conceivably, you could end up with something that functions, but it would be far more reasonable to simply build the diesel in the first place. Biologically speaking, however, what is efficient from a constructive point of view is not what we have, and what we have makes sense in terms of evolution. Fish preceded mammals by around 200 million years. As fish spread through the marine and aquatic niches available to them, some ventured into shallow areas where they were exposed to air for periods of time. Those that could survive in that environment prospered.Over time, modification of their existing structures allowed taking better advantage of this new, oxygen-rich resource that served them well.

When organisms adapt to new environments over time, and existing structures become modified to serve newer functions, sometimes some older structures become superfluous and may be discarded, again over time. There have been a number of passionate arguments about this, as sometimes an organ that appears to be in the process of undergoing evolutionary discard may continue to serve an

B. Marcus, *Evolution That Anyone Can Understand*,
SpringerBriefs in Evolutionary Biology, DOI: 10.1007/978-1-4419-6126-6_4,

important purpose. Sometimes, however, the organ being discarded remains as what amounts to a useless relic. One case particular to humans is the human tail.

Somewhere in the second month of our prenatal development, our primitive vertebral column, spine if you prefer, projects beyond the anus as a short tail. It never amounts to much, and it regresses rather quickly. At birth, the only remnant of our tail is a short projection of a few fused bones that extends minutely beyond the bone that attaches the backbone to the hips. We call this projection the coccyx. Some anatomists have argued that the coccyx represents the vestige of the tail possessed by our tailed ancestors. Thus, it has been described as a vestigial organ, a remnant, or vestige of a larger structure that existed during our biological history. Some creationists deny that the embryonic structure ever exists, begging the question of what the post-anal projection happens to be. Others claim that since there is a muscle attached to the coccyx, it has a function and cannot be vestigial, and indeed they are partly correct. The muscle that attaches to the coccyx, the *extensor coccyges*, is, in tailed animals, the muscle that moves the tail.

Anatomically, vestigial organs are remnants of larger organs of our biological past, but that does not mean they are now without function. The human appendix is one example. It is a small, tubular extension of the large intestine that dips down from the large intestine near its junction with the small intestine. It occupies the same anatomical location as the cecum of animals like the rabbit, pig, and horse, in which it is a much larger organ that functions as an additional digestion site for the coarse vegetable material that those animals eat. Many humans function quite well without an appendix, which has perhaps led to the possible misconception that it plays no role and indeed because it is sometimes prone to inflammation, one is better off without it. Recently, however, scientists have determined that the appendix is richly supplied with lymphoid tissue, which plays a role in immunity. Moreover, other scientists have determined that the appendix may serve as a reservoir for the beneficial bacteria that inhabit our large intestines and help in our digestion of vegetable fiber. Following severe diarrhea or some disease that causes it, bacteria from the appendix may repopulate the large intestine.

Another organ long thought to be vestigial is the thymus gland, which is located in the chest cavity just above the heart. Prior to birth it is a fairly large organ, extending from the base of the throat to the heart. However, it appears not to grow at all after birth, which in the past led to the belief that it had no function. That it does have a function, a rather important one in the development and activity of our immune system, began being understood within the last 50 years. Such discoveries force scientists to re-evaluate their positions on a particular issue, but it does not force them to re-evaluate their positions on an overall subject, as those who oppose evolution would argue. The elucidation of functions of the thymus and the appendix does not refute that the organs in humans are vestiges of what they are in other animals and early in life, respectively, and there are other vestigial structures that continue to be functionless, from genes on up.

University of Wisconsin biologist Sean B. Carroll tells the story of the Antarctic icefish, a bottom-dwelling fish that lives in the super-cold waters of the Antarctic Ocean and, among other odd characteristics, has white blood. The icefish produces neither hemoglobin nor red cells, the chemical that carries oxygen in vertebrate blood and the cells in which it is found, respectively. Living in the oxygen-rich water of the southern ocean, it is able to absorb the oxygen it needs, and its large heart is able to successfully deliver the oxygen to its tissues. But within the icefish's DNA there are vestiges of genes for hemoglobin. Similarly, genes associated with chlorophyll and photosynthesis, the material and process by which green plants make their food from sunlight and raw materials, have been found in the cells of the organism that causes human malaria, a parasite that never sees sunlight and has no need to manufacture food. It lives within its food supply, a rich one at that. It never has to make food from raw materials. Indeed, it is incapable of doing so. Yet it has vestiges of genes for that very purpose. Even humans have vestigial genes. One example would be those associated with our sense of smell. We might be described as olfactorily challenged when compared to other animals. However, we still have a collection of no longer functioning genes, located on almost every one of our chromosomes that are associated with our anemic sense of smell. These are likely to be remnants of the genes that functioned in our ancestors.

It is well known that the sense of smell is vitally important to most mammals as their principal means of acquiring information about their environments, and humans are a conspicuous exception. Even though we have adequate genes for sharp olfactory reception, smell, on our chromosomes, roughly 60% of them do not function. In apes, our nearest biological relatives, the number is more likely 30%, and in animals like dogs and mice, it is much smaller. On the other hand, humans and most apes have full color vision. Most other mammals do not.

The usual interpretation of this is that when full color vision evolved in the common ancestor of humans and apes, those organisms came to rely more on it and less on smell for environmental information. With no need for an acute sense of smell, the loss of gene function for smell did not cause any disadvantage. As those genes lost their function, our ancestral organisms became carriers of vestigial genes. As our more recent human ancestors ceased to be arboreal, that is they became ground dwellers rather than living in trees, and evolved erect posture, they came to rely on vision even more and smell less. As a result of this, today we continue to carry much DNA devoted to olfaction, but most of it is in nonfunctioning, vestigial genes.

Proponents of divine creation and intelligent design argue that all forms of life were put together more or less as they now appear. In other words, icefish were made without red blood cells, humans were made with a poor sense of smell, and hemoglobin and malaria parasites were made with photosynthetic material. If these are indeed the case, one would be compelled to ask why so elegant a creator or designer would have burdened these creatures with worthless molecular machinery. Using again the analogy of the railroad locomotives, it would be much like building a non-functional steam boiler into a diesel engine.

Fossil genes open another discussion on the ideas of design vs. evolution. Genes are composed of DNA, deoxyribose nucleic acid, a fact that is now pretty much common knowledge. What may not be common knowledge, however, is how little of our DNA actually functions as genes. Genes are defined as sequences of DNA that code for protein. This means that a section of a DNA molecule in the nucleus of a cell sends a chemical message into the surrounding fluid of the cell. This message is used more or less as a blueprint for constructing protein. A case in point would be the protein hormone insulin that is made in the pancreas. Genes dictate the production of insulin by the pancreas cells. However, less than five percent of all human DNA, for example, codes for protein. The rest is noncoding. Indeed, most multicellular animals and plants contain such DNA. Some of it is the vestiges of once functional genes, but much of it appears to have come about from repetitive reproduction of shorter regions of noncoding DNA.

Some noncoding DNA appears to function in promoting the activities of genes, which has prompted some design proponents to argue that it was put there deliberately. However, most noncoding DNA does not serve that function. Instead, ironically, it may serve as a raw material for evolution, as some of it can be moved around and by being inserted into DNA sequences may form new genes or modify existing ones.

The icefish has a number of characteristics that make it well adapted to life in its extreme environment. It is a bottom dweller, and it lacks a swim bladder, making it unable to adjust its vertical position simply by adjusting its body density as most fish can. However, it has a very light skeleton so that if it chooses to swim upward, it can do so without being weighted down by its own bones. In addition, its tissues contain a natural antifreeze made out of a glycopeptide, a protein–carbohydrate complex. The group of fish into which the icefish belongs, the Notothenioiei, dominates the fishes of the Antarctic waters, and tissue antifreeze is common among them, but only the icefish species lack hemoglobin and have light bones.

Notothenioiei in general are found in the cold waters of the southern ocean, and many live at temperatures near, at, or below the freezing point of fresh water. Because of its salt content, seawater can be cooled to less than 0° C (32° F) without freezing, but ice crystals can form in the tissues of fish in such water. Because most fish cannot regulate their internal body temperature to any great extent, they are usually as warm or cold as the water around them. Notothenioiei get around the problem of living in sub-freezing water by producing an internal antifreeze made out of a glycopeptide, a combination of short protein molecules combined with a carbohydrate. One can visualize that the mutations dictating the production of antifreeze occurred before some species of the Notothenioiei were restricted to sub-freezing waters. The presence of the antifreeze, however, allowed their venturing into such waters, where they found little or no competition, and it further allowed their survival in that cold water. Since oxygen dissolves more readily in colder than warmer waters, cold water contains higher concentrations of the gas, and the loss of hemoglobin proved to be no great disadvantage. Thus, we now have bloodless fish, so to speak, living in subfreezing water.

Another animal that must contend with sub-freezing conditions is the North American wood frog. This is the only frog that lives in north of the Arctic Circle; it escapes the cold winters by hibernating, but it does not always succeed in avoiding the cold winter temperatures. It too produces antifreeze, but unlike the Notothenioiei, the frog uses glycerin and glucose as antifreeze. Thus, it solves the same problem the icefish faces, but it uses different chemicals and therefore different genes to do so.

That two essentially unrelated animals would use different chemicals to solve a similar problem is perfectly consistent with evolution theory. Evolution favors whatever works. How it may occur is not an issue. That two different but equally effective solutions would have been used by a single designer is illogical and unnecessarily complicated. Moreover, in the case of the icefish, a designer would have placed these fish in a biological corner. Right now, temperatures in the southern ocean are increasing, and evidence suggests that they will continue to do so. The problem of global warming is affecting the oceans as well as the atmosphere, and there is no reason to not believe that eventually the Antarctic Ocean will warm above freezing temperatures for at least short periods of time. Once that happens, their lack of hemoglobin and consequent inability to bind oxygen will condemn the icefish to extinction.

Design proponents have argued that because some noncoding DNA has been shown to have a function, all of it must, and it must have been put there deliberately. Although the logic is flawed, until it is definitively shown that at least some noncoding DNA has no function, one cannot categorically declare them entirely wrong. However, while design proponents can satisfy themselves with arguing for the design of noncoding DNA, what can they say about hipbones in a snake or a whale?

Hipbones, of course, are the means by which hind legs attach to the bodies of terrestrial vertebrate animals. Snakes do not have legs and are well adapted for a life without them. Their long, tubular bodies allow them to slither along the ground quite easily, travel through burrows underground, swim, and even climb trees. Indeed, they conserve space within their bodies by having only one functional lung, the other being only a vestigial remnant. But among the constrictors, the grouping of snakes that includes boas, pythons, and king snakes and is recognized by biologists as primitive, relect hipbones are present. Moreover, a fossil of a snake named *Najash rionegrina*, found in Argentina, had not only hips but also legs as well, supporting the idea that snakes evolved from lizard-like ancestors (Groshong K 2006). Parenthetically, leglessness is not confined to snakes. Even today, there are lizards that are legless or nearly legless, and there are legless amphibians as well.

Whales too often contain vestigial hipbones. The baleen whales like the wright whale and blue whale typically have well-formed hipbones, despite there being no obvious function for them.

Fossil or vestigial structures occur because organisms have genes dictating their development. The genes remain functional even though the structures do not. So whales have hip bones because they have genes that dictate the formation of hipbones even though genes dictating the formation of legs are either missing or nonfunctioning just as humans have genes dictating the production of tail-wagging muscle despite the absence of a tail. It is possible that one day whales and constrictors will lose their hipbones. The genes controlling their development could conceivably erode and become like the fossil genes for hemoglobin in icefish. Nobody can say with any certainty whether or not this will occur, but it is conceivable because loss of those structures or the genes dictating their presence will not put the organism at a disadvantage. A whale could lose its hipbones and probably never notice that they were missing. Indeed, it probably has no idea that it even has any.

Recently some scientists have become reluctant to use the term vestigial because of discoveries of function in organs once thought to be functionless, particularly in humans. The thymus mentioned earlier is one example. However, the term applies to more than simply organs. For example, why do we sometimes sneer when we show contempt or hostility? It has been argued that the expression is a behavioral response akin to an animal baring its teeth, a threat ,or dominance display. But how likely are we to actually bite the person at whom we are sneering? The response is vestigial. Furthermore, why do we retain muscles to wag our tails and to move our ears for that matter? Admittedly they are unimpressive in comparison to those animals that actually can move their ears and have tails to wag, but they exist, without function. Finally, what is the function of goose bumps? They occur when we get chilled but serve no purpose to warm us. In furry animals, the same response fluffs the fur and traps warm air. In us it appears to be a completely vestigial response.

Vestigial structures abound in the animal kingdom. Ratite birds, such as ostriches, retain wings that are useless as organs of flight. Among kiwis, the wing remnants are so small that they are not visible unless one parts the feathers over them. Indeed, they're too small to serve any function, including as agents of balance, which the wings of ostriches could conceivably do. Such excess baggage and the genes that dictate their development, as well as our fossil genes and useless gestures support the idea that our distant ancestors had use of them. Many species have abandoned uses of structures or perhaps use the structures toward different ends than their ancestors did. Whatever the case may be, it backs the notion of change over time, and that is basically evolution.

Reference

1. Groshong K (2006) Oldest snake fossil shows a bit of leg. New sci 18:58 http://www.newscientist.com/article/dn9020-oldest-snake-fossil-shows-a-bit-of-leg.html. Accessed 24 June 2011

Chapter 5
Islands in the Sky and Elsewhere

The Monteverde Cloud Forest of Costa Rica is considered to be one of the Western Hemisphere's premiere reserves of biodiversity. It sits around 10^0 north of the equator and near the continental divide at an elevation above 1,000 m (3,300 feet), where it is often enshrouded in clouds. The cloudwater condenses on the trees, rocks, and anything else that happens to be present, supplying vast quantities of water to the forest. There one can find around 50 species of hummingbirds in addition to perhaps another 350 bird species, plus tree ferns, orchids, bromeliads, butterflies, and amphibians. Among the last category was a small, iridescently bright yellow toad. The so-called golden toad of Monteverde used to be found in the montane forest of the region. Sadly, it became extinct around 20 years ago. Gold-colored frogs do occur elsewhere on the planet, but the specific species *Bufo periglenes*, the Monteverde golden toad, exists nowhere else, not even on adjacent mountains in Costa Rica.

In a biological sense, mountaintops are like islands, thus the title of this chapter. Instead of being surrounded by water, they are surrounded by a different environment that may be every bit as hostile to the mountaintop-dwelling organisms as water is hostile to the land-dwelling organisms that live on true islands. Indeed, any isolated environment is an island, metaphorically if not actually. As mentioned in the first chapter, islands often harbor unique varieties of plants or animals, and the same is true of mountaintops or any other kind of island, metaphorical or real.

Speciation often occurs when a continuously interbreeding population of organisms somehow gets fragmented, and a part of the population becomes reproductively isolated from the rest of it, as if it was suddenly removed to an island. The isolated fragment can diverge until it becomes sufficiently different from the parent population that it is no longer capable of successfully breeding with the parent population. This, by definition, classifies it as a separate species. The process can be facilitated by the isolated fragment of the population being somewhat genetically different than the parent population. This is often the case because animals that inhabit a broad range of environments face different

B. Marcus, *Evolution That Anyone Can Understand*,
SpringerBriefs in Evolutionary Biology, DOI: 10.1007/978-1-4419-6126-6_5,

challenges in some parts of the range than in others. Individuals at one edge often differ from those in the core or at other edges.

To understand the reasons behind this, consider the problem of temperature maintenance in warm-blooded animals like black bears. Like ourselves, black bears when they are not hibernating maintain a more or less constant internal body temperature. In the case of those living in central Mexico, for example, a problem they may face is having the need to shed excess heat. In contrast, those living in northern Alaska often face the problem of conserving heat, even when they are not hibernating. Ecologists have noted that warm-blooded animals in colder parts of their range often tend to be larger in size, lighter in color, and possessed of smaller extremities than those in the warmer regions of the range. Indeed, they speak of Bergman's, Gloger's, and Allen's rules, respectively, as the principles describing these phenomena. Consequently, one would expect that Mexican black bears would be, indeed, black, with relatively small bodies and, say, large ears. In contrast, larger bodied, smaller eared, lighter colored bears would more likely be found farther north.

Fur color in black bears is highly variable; a number of color variants exist. Conventional wisdom has it that their color allows them to blend in with their surroundings. Consequently, it would make sense that within a dense forest where a bear is constantly in shadows, black fur would allow it to approach its prey unseen. The problem with that logic, however, is that bears do not sneak up on prey that often. Unlike other carnivores like wolves and large cats, bears, though quite willing meat eaters, are not highly predacious. They are generally omnivores more than total carnivores, eating virtually anything that is easily available, including insects and plants. And while they are perfectly capable of killing and eating other animals when they really want meat, their preference is to eat something that died some other way. In other words, they are more scavenger than predator. It takes much more energy to chase and kill something than it does to eat something that died some other way, and conserving energy is often important to an animal's health and well being. Moreover, even when bears do kill, their prey are more likely to rely on scent than sight for environmental information, and so it would be more important for the bear to approach from downwind than it would be to sneak up unseen. This is not to say that camouflage plays no role in the natural selection of a bear's coat color, but it may not be the major factor.

As indicated earlier, bears are warm-blooded animals. Essentially, this means that they generate heat internally, which they rely on for body warmth. They typically do not absorb heat from their environment as does, for example, a frog. When the air is cold, a bear's metabolism will speed up to generate more heat, which is then trapped by its fur, much as the heat we generate on a cold day is trapped by clothing. In warmer weather, a bear's metabolism would slow down to generate less heat, and its fur would insulate it against external warmth. However, on a warm day, becoming overheated could be a problem if a bear was not able to shed excess heat. Here a darker colored animal would be at an advantage, according to Gloger's rule, because light colors tend to radiate less heat than darker ones, as one can easily demonstrate were he to step on a black, macadam driveway on a hot, sunny day. One would find it warm to the touch. In contrast, a lighter colored, concrete driveway would be cooler. Similarly, a black bear in Mexico would shed heat more efficiently than a lighter one, and the light one would conserve heat better in Alaska than a dark one.

Color variants in black bears often occur in clusters, indicating that genes dictating coat color occur in clusters. A case in point is the so-called Kermode or spirit bears that occur on some small islands off the west coast of British Columbia, Canada. The water separating the islands is not wide, and it is probable that all of them were part of a single, larger island or possibly even the mainland during the last ice age when sea levels were lower. Some of the bears of these islands are white in color. They are not albino; their eyes and noses are normally pigmented, not pink as would be the case in albinos. However, they result from the same sort of genetic process: a mutation in the color-determining genes that changes the more typical distribution of pigment in the animal. Within the Kermode bear's range, the mutation did not put the individuals who expressed it at any particular disadvantage, and so it survived. Indeed, during the past glacial period, it may even have been advantageous, given Gloger's rule. Oxymoronic sounding white black bears have been seen elsewhere within North America, suggesting that the mutation or one like it has occurred in other places, but only rarely. Only within the Kermode bear's range are the mutants at all common.

Another light colored variant of black bears is the so-called glacier bear that is found in the northern part of the Alaskan panhandle, around heavily glaciated Glacier Bay, and into nearby Canada. This animal has a white or yellowish outer coat and a blue or gray inner coat. It most likely originated in an isolated population, again on a metaphorical island in effect, and resulted from a gene mutation or mutations that were at worst not harmful and at best beneficial. The bear is difficult to see on ice giving credence, in this case, to the argument that camouflage played a role in the selection and preservation of this mutation.

One mechanism that contributes to why genetic variants tend to show up clustered in small, often isolated fragments of larger population, such as one would find on an island, is a phenomenon called genetic drift. This is a random process where a particular mutation becomes abundant in a population for no reason other than the luck of the individuals who carry it. They survive in large enough numbers to maintain the gene. In a small enough population, it can alter the general appearance of its members. For example, water snakes on the islands of western Lake Erie routinely differ in color from those on the mainland. Genes affecting behavior, body size and structure, strength, speed, and factors in appearance in addition to color can also accumulate randomly in small populations and population fragments as a result of isolation and drift. In time, the parent and daughter populations can become sufficiently different from one another that members of each do not recognize the other as potential mates. When they are no longer capable of breeding with one another, speciation can be said to have occurred.

From a biological perspective, any kind of environment that is completely surrounded by a different kind of environment can be thought of as an island, as indicated earlier. Thus, a stand of trees surrounded by prairie is akin to an island, as would be a meadow surrounded by forest. Some plants or animals in one of the surrounded environments may be incapable of crossing the surrounding environment and thus are genetically locked in, much as is a tortoise on one of the Galapagos Islands. Such organisms isolated on these islands experience gene mutations that cannot be shared with members of the population with which the segregated individuals were once continuous, nor have they access to gene mutations that occur in the parent population. Moreover, different selective pressures within the island and the randomness of genetic drift mean that the island population eventually diverges from the parent population. Given enough time, the differences can become substantial. Thus, for example, the island of Madagascar off the east coast of Africa has been isolated long enough that three quarters of its native species are unique to the island.

True islands, bits of land surrounded by water, usually form as a result of geologic events, such as underwater volcanoes, as is the case with the Hawaiian Islands and Iceland, for example, or by the fall and rise of ocean levels as ice ages come and go, as is the case with the British Isles. Exceptions exist, such as Barro Colorado Island, which resulted from a mountain peak that was isolated when Gatun Lake was formed during the construction of the Panama Canal. Metaphorical islands like mountain tops can also result from geologic processes. However, many metaphorical islands as well as some real ones like Barro Colorado are the result of human activity. For example, rain forest clearing, such as is going on in much of the tropics at this time, often results in the destruction of vast acreages of trees, but fragments of the forest are often left behind. Such fragments of intact forest that are surrounded by destruction or by farmland that often results from clear cutting forests are consistent with the concept of islands thus far developed. Many species of forest animals will not cross open territory and are thus isolated within the forest islands. Sometimes such islands are deliberately left behind in the name of conservation as parks.

A reasonable question at this point might be: how likely is it that a fragment of a larger population may become isolated on a real or virtual island? The answer is very! In the case of virtual islands resulting from human activity, remnants of previously larger populations can easily become isolated, as for example, howler monkeys and coatis that were trapped on Barro Colorado Island when Gatun Lake filled. Population fragments do not always survive. Likewise on Barro Colorado Island, pumas trapped on the island could not find adequate prey and became extinct. However, even barren islands can become populated, or in some cases, repopulated.

On August 27, 1883, the island of Krakatoa in the Sunda Strait of Indonesia exploded. The island was volcanic, and the volcano erupted with a force estimated to be in excess of 10,000 times that of the bomb that was dropped on Hiroshima. The sound of the explosion could be heard thousands of miles away, giant tsunamis curled throughout the region and beyond, and the island was essentially, if not thoroughly, sterilized. Reportedly, the first visitors to the island in the following year found a single spider, possibly blown to the island by the wind, and nothing more. However, in time the island was repopulated with plants and animals.

Animals like bats and birds usually manage to get to islands on their own. Sometimes a storm helps. Often these animals may carry seeds in their guts, seeds that they may leave on the island when they defecate. When one of these seeds falls onto a suitable surface, it germinates and a plant grows. Other animals, and possibly plants as well, may travel to islands by a phenomenon called natural rafting, where an organism is caught on some kind of floating debris. This could happen in the case of an arboreal lizard or snake that was on a tree that fell into a body of water and then floated out to sea and washed up on an island. Likewise, animals aboard coastal vegetation that was ripped from its original location by a storm surge may find themselves on route to an island. This may sound like a stretch, but animals have been seen on flotsam out at sea. Perhaps it does not occur often, but what may be a rare event in terms of a single person's experience can be rather common when considered in terms of decades or centuries.

Yet another way of animals getting to islands is by swimming. Within a few years, reptiles had returned to Krakatoa, probably from the nearby islands of Java and Sumatra. It is likely that many of them got there by swimming.

In recent history, relatively speaking, the task of island hopping has been made significantly easier by human activity. Cats and dogs have been distributed around the world by humans. Rats have been following humans around the planet throughout the ages. When humans took to sea, rats often came along on the boats as uninvited guests, or perhaps even invited ones. It has been said that the Polynesians ate rats, thus they often carried them along on their sea-going ventures. In some instances, rats caused problems on islands to which they were transported. Humans than introduced animals like the mongoose to control the rats. And so, the flora and fauna that exist on islands is now much different than it probably would have been were humans not part of the story.

Once animals get to islands, by whatever means, the isolation leads to changes in their appearance that most likely would never had occurred had they remained in their original habitat. A case in point might be a rather odd bird that was once found on the Island of Mauritius in the Indian Ocean, east of the Island of Madagascar: the dodo.

The dodo, extinct now for three centuries, was a large, flightless bird that existed only on Mauritius. It had lived there for perhaps eight million years. Descended from the pigeon, the dodo's ancestors found a beneficial environment on Mauritius when they first arrived, probably blown there by a storm. There were no predators, and there was an abundance of food. Consequently, gene mutations that hindered the bird's ability to fly proved to be no disadvantage, and over time flight was lost completely. It was not necessary for the birds' survival and so it was biologically abandoned. Moreover, by the time Mauritius was discovered, around 1,600 by Portuguese sailors, the dodo had become quite large, perhaps a meter more or less in length and 15–20 kg (30–45 pounds) in weight. Conventional wisdom has it that the dodo became extinct because it was too stupid to run from the sailors, who slaughtered it for food. More recently, however, scientists have come to believe that it was the introduction of domestic animals like cats and pigs plus the ubiquitous rats that accompanied human explorers that caused dodo extinction by destroying the animals' nests. Moreover, clearing of forests on the island robbed the dodo of food. The dodo was well adapted to its environment, and its extinction was probably more a case of its environment being altered too quickly for it to adjust than it was a matter of the birds' intelligence or lack thereof. The phenomenon is known as habitat destruction, by the way, and it has been responsible for the extinctions of many species of animals and plants.

Population fragments of species that become isolated on islands may end up differing from the parental population in a number of ways. The Nēnē, a goose that is confined to the Hawaiian Islands, is similar to the Canada goose with whom it probably shares a common ancestor. It adapted to its island habitat a bit differently than the dodo; it evolved into a smaller version than its sibling species, but it too is flightless or nearly so having no need for flight as do other geese. The webbing in its feet is also reduced as compared to other geese as the Nēnē is more terrestrial than they are. The Nēnē most likely descended from a population of migratory geese that were blown to Hawaii during a storm. Finding a suitable environment, they were able to survive well until human interference, hunting for example, became a factor in their survival. The Nēnē is now endangered, although efforts are being made to conserve the animal. However many times geese were blown to the Hawaiian Islands by storms can never be known, but it is likely small in number. It is probably not all that unusual for geese to have been blown off course and over the Pacific Ocean during migrations, but the probability is that most of those that had the experience died as a result.

There are more examples of island animals that are related to mainland species but differ in one or more ways, and in many cases, the difference is a matter of size. Island species are often either extremely large or extremely small specimens of their particular grouping. This particular phenomenon was described by biologist J. Bristol Foster and is sometimes known as Foster's rule.

One case in point is the Komodo dragon, a monitor lizard that is found on a few islands of Malaysia, east of Java. Monitor lizards in general are found throughout much of the tropical old world, from Africa through southern and southeastern Asia, islands of the Indian Ocean and South China Sea, and Australia. They range greatly is size, but the Komodo dragon is the largest of the category. Reaching a length of over 3 m (10 feet) and a weight in excess of 130 kg (300 pounds), it is clearly the giant of the group. By the same token, the giant tortoises of the world are island species occurring in places as remote from one another as the Galapagos Islands, Madagascar, Indian Ocean Islands, and Malaysia. Many of these animals are now extinct, but it is likely that as species, at least some arose independently of the others. Conceivably, the species that live around the Indian Ocean could have descended from a common ancestor, but the Galapagos tortoises most likely are the successors of a tortoise from South America that floated to the archipelago.

The other side of Foster's rule is a phenomenon called insular dwarfism, something that is more likely to occur among mammals. In this case, animals on islands tend to be smaller than those on the mainland. Examples include extinct dwarf elephants that were once found on Mediterranean Islands like Cypress and Malta as well as some from the Malaysian and other Asian islands. Many other illustrations exist, including a small tiger that lived on the island of Bali and became extinct perhaps less than 100 years ago. There is even a small human, the so-called Indonesian Hobbit, that once lived on the Indonesian island of Flores, but that belongs in another chapter.

Perhaps the most ironic case of the kind of evolution that occurs on islands involves fish in bodies of water isolated in a desert. Specifically, Death Valley in Southern California, described as the hottest locality in North America, is home to several species of unique fish. Known collectively as desert pupfish, these are small killifish that appear to have descended from a common ancestor. Death Valley has not always been desert. Indeed, during the Pleistocene epoch, the region was covered by a large lake in which the ancestor of the desert pupfish lived. As the lake dried, pools of water became isolated, trapping and separating fragments of the ancestral pupfish population. With time, these isolated remnants of that ancestral population evolved into the several species of desert pupfish that now exist, often confined to a single spring or pool just as the species of Darwin's finches each occupy a separate island in the Galapagos.

Some of these species are barely hanging on. Death Valley continues to grow warmer, and for some time, groundwater beneath Death Valley has been removed for agriculture. At least one spring harboring pupfish was impacted. It is likely that within the foreseeable future, one or more of the habitats that currently house pupfish will dry completely. If so, the species will disappear unless specimens are removed to a new habitat. Indeed, that has been tried, but in their new habitat with different environmental conditions, the species changed (Lema 2008).

The isolated water holes that pupfish occupy in Death Valley could conceivably be described as oases. An oasis can equally conceivably be described as an island of water surrounded by a sea of hot, dry sand. It is yet another instance of a particular environment being surrounded by one that is completely different and totally inhospitable to the environment's inhabitants. In the case of oases, avenues for migration are essentially nonexistent, and whatever inhabitants they house are left to exchange genes only among themselves.

The phenomenon of isolation on islands is even further illustrated by islands within islands. Stephen Jay Gould described volcanic islands in the South Pacific, where valleys were separated from one another by volcanic ridges. Land snails that had somehow found their way to the islands established themselves in different valleys and evolved into separate species (Gould 1993).

Whenever a population becomes so isolated that genes from related outside populations cannot enter, a condition called inbreeding occurs. Often the results of inbreeding are pathological, such as the problems that routinely occur among purebred strains of animals. It even crops up among humans, for example the Tay Sachs disease that is sometimes found, independently, among Eastern European Jews and French Canadians. Islands, whether they are the conventional kind that are surrounded by water or the metaphorical kind like a mountain top can be so remote that gene flow into and out of animal populations living on them is severely restricted, and mutations that are rare in more extensive populations of the same species can become locally abundant. Such changes in genetic content of the population ultimately result in changes in the appearance, behavior, or some other characteristic(s) of the island population. Consequently, animals of the same or closely related species may differ markedly from island to island and from a mainland species from which they all descended. Some species may be unique to a single island. Combinations of different mutations and selective pressures, along with the randomness of spontaneous mutation and genetic drift, make island populations of animals truly singular. Consequently, islands have been described as evolution laboratories, and as more and more ecological harm occurs on islands, again whether actual or metaphorical, the loss of the genetically unique species that populate them, such as the golden toad of Monteverde, can only increase.

References

Gould SJ (1993) Unenchanted evening. In: Gould SJ (ed) Eight little piggies: reflections in natural history. Norton, New York, pp 23–40

Lema SC (2008) The phenotypic plasticity of Death Valley's pupfish. Am Sci 96(1):28–36

Chapter 6
Superbugs

In 1928, a Scottish scientist by the name of Alexander Fleming discovered that a blue–green colored mold had contaminated a culture of *Staphyloccus aureus* bacteria he had been growing and appeared to be killing them. As things turned out, the discovery could be described as serendipitous, because Fleming had been working on something else at the time, and the contamination occurred while he was out of the laboratory.

As one story goes, Fleming was investigating the antibacterial effect of nasal secretions. He had inoculated a culture dish with the bacteria, but rather than place it in an incubator, he left it on his laboratory bench and went on a 2-week vacation. During his absence, a mold spore found the culture and was able to grow. The bacteria grew as well, but owing to the colder temperature of the laboratory bench, not as quickly as they would have in the incubator. When Fleming returned, he found a clear zone of no bacterial growth surrounding the mold. He realized that somehow the mold was preventing bacterial growth.

The mold, *Penicillium notatum,* belongs to a genus that is commonly found in the soil and opportunistically grows on vegetable material. It can cause the molding of fruits, for example. Fleming could have easily discarded the contaminated culture dish, as had earlier investigators who had had similar experiences. Instead, he chose to further investigate the phenomenon and as a result learned that the antibacterial material in the mold, which he called Penicillin, could kill bacteria in mice without harming the mice. Regrettably, he was never able to provide Penicillin in a medically useful form, but in 1941, two English scientists, Howard Florey and Ernst Chain, accomplished that.

While Fleming's discovery of Penicillin can be described as serendipitous, and versions of the story of his discovery other than the one presented here exist, one must remember that it was his powers of observation, intellectual curiosity, and subsequent experimentation that ultimately led to the conclusion that he had a potentially useful product on his hands. In essence, serendipity favors a prepared mind.

B. Marcus, *Evolution That Anyone Can Understand*,
SpringerBriefs in Evolutionary Biology, DOI: 10.1007/978-1-4419-6126-6_6,

Another category of drugs discovered more or less serendipitously was the sulfonamides or sulfa drugs. These were originally developed as dyes and were found to be antibacterial. The discovery of more recent antibiotics has resulted from deliberate searches, the result of which, along with Penicillin and Sulfa, brought about the control of many infectious illnesses. Indeed, the use of antibiotics during World War II may have been the principal reason for its being the first war in history in which combatants who survived wounds suffered in battle were not condemned to die of subsequent infection. Moreover, long-time killers like pneumonia and tuberculoses were brought under control by antibiotics. More recently, however, after years of use, antibiotics are losing their effectiveness, and many infectious bacteria that were once stopped cold by antibiotic therapy are no longer touched by it.

A story somewhat similar in result can be told about chemical insecticides, particularly the compound dichloro-diphenyl-trichloroethane, more commonly known as DDT. The compound was first synthesized in 1874 by Othmar Zeidler, a doctoral student at The University of Strasbourg in Alsace in France, but its original use, if any, seems to have disappeared into history. In 1939 its insecticidal properties were demonstrated by the Swiss chemist Paul Hermann Müller, who was later awarded a Nobel Prize for his work. DDT appeared to be as miraculous in killing insects and other arthropods, including disease vectors like ticks and mosquitoes, as Penicillin was in killing bacteria. It too contributed to the reduction of infection during World War II by eliminating typhus-carrying ticks in Europe and Malaria-carrying mosquitoes in the South Pacific. Indeed, countless lives have been saved by DDT and by Penicillin, as well as other antibiotics and insecticides, most of which continue to be used today, even though in many cases their effectiveness has waned considerably. Ironically, in saving so many lives, pesticides and antibiotics may have contributed to the explosive growth in the human population that is now occurring and is resulting in crowding and malnutrition that promote disease.

Both antibiotics and pesticides can be described as chemical weapons against what are often colloquially referred to as "bugs." And in both cases, the loss of effectiveness has nothing to do with the compounds; it has resulted from genetic changes in the target organisms, the bugs.

To understand how this happened, consider our own species and its resistance to disease germs, influenza for example. Somebody with the flu gets on an airplane and for the next two hours coughs and sneezes into the sealed passenger compartment. Everybody else on the plane is going to be exposed to that person's germs. However, not everybody will come down with the flu. Of those that do, some might develop a mild case, some might get one that is more serious, and possibly one or two might develop symptoms severe enough to develop into pneumonia or even cause death. In other words, that particular population shows diversity in its susceptibility to influenza. By the same token, individual members of a population of bacteria that is susceptible to Penicillin, using the most appropriate example, will vary in their susceptibility. A given dose of the drug may kill 99% of the population; the one percent that survives is the most resistant.

Moreover, their resistance is genetic, and when they reproduce, they pass their resistance on to their descendants, more of whom will survive subsequent treatments with the drug. Consequently, the Penicillin can work as a selective agent, eliminating the susceptible individuals from the population and choosing the resistant ones for survival. In time, it can force the evolution of a population of bacteria from largely Penicillin susceptible to largely Penicillin resistant. The same mechanism is true of other antibiotics as well.

The mechanism also applies to DDT resistance in arthropods, particularly mosquitoes. When first exposed to DDT, mosquitoes varied in their susceptibility to the chemical just as bacteria varied in their susceptibility to antibiotics. Some may even have been totally resistant. Consequently, rather than eliminating all mosquitoes, DDT applications eliminated the most susceptible ones only. Selective elimination of the most susceptible ones left the most resistant to survive and pass on their resistance. Admittedly this was a small minority of the insects, but it was enough to reproduce and pass on resistance. The cycle of treatment and survival was repeated over many years, but eventually, in many parts of the world, mosquito populations evolved from DDT susceptible to DDT resistant.

There are perhaps several reasons why bugs of both types have evolved resistance to their respective controlling agent. One is that these organisms, bacteria and mosquitoes, breed very rapidly. They are capable of producing multiple generations in a short period of time, which provides opportunity for both abundant gene mutation and rapid genetic turnover. Longer lived, or perhaps more appropriately, slower reproducing organisms would not respond nearly as quickly. Moreover, many bacteria that can cause infection can also live in soil, where they are routinely exposed to fungi. Many fungi produce antibacterial compounds, and bacteria have been evolving resistance to those compounds essentially for as long as they have been on the planet. More importantly, however, is the fact that antibiotics and DDT have been misused and overused.

Antibiotics in general work only against bacteria and not always against all bacteria. Bacteria are often characterized according to the way they react to a staining technique invented by a Danish scientist by the name of Hans Gram in 1882. Consequently, the technique is now known as Gram's stain. Bacteria that react in a particular manner are described as Gram positive, while those that do not are Gram negative. Penicillin historically worked best against Gram-positive bacteria, such as *Staphylococcus* and *Streptococcus*, organisms that are often responsible for skin and respiratory infections, respectively. In contrast, a Gram-negative infection like *Salmonella* poisoning or typhoid fever would not be as readily affected by Penicillin. However, that has not always stopped physicians from trying it. No antibiotic works against viral infections like the flu, the common cold, or measles. Again, however, physicians have often prescribed antibiotics for such ailments. In fairness, it was often in an effort to prevent subsequent bacterial infection, or it was because patients pressured the physician for an antibiotic. The result, however, was the further exposure of bacteria to antibiotics and the consequent elimination of susceptible individuals and selected survival of resistant ones.

In other cases, antibiotics were prescribed appropriately but unnecessarily. For example, earaches are a common malady in children, and pediatricians often prescribe antibiotics to treat them. Earaches are indeed sometimes caused by bacterial infection, and often a course of antibiotics is proper if the problem fails to clear up on its own. However, earaches often do clear up on their own without antibiotics. By the same token, antibiotics have routinely been prescribed to treat sinus infections, which also can clear up on their own. It is often a judgment call, and many physicians prefer to err on the side of caution. That notwithstanding, the result was often a selective elimination of susceptible bacteria, and we are now faced with a serious problem in terms of bacterial drug resistance.

One of the best places to pick up a bacterial infection is, ironically, in a hospital, which is often a breeding ground for antibiotic resistant bacteria. Moreover, people in hospitals are often immune compromised and are consequently vulnerable to infection. Of particular concern are the so-called methicillin-resistant *Staphylococcus aureus* bacteria, technically known as MRSA. *S. aureus* is a skin bacterium; everyone carries it around on his body surface. MRSA bacteria are resistant to most, perhaps in some case all, antibiotics. The bacteria have been exposed to so many antibiotics for so long that the susceptible individuals have long since been selected out of existence and have been completely replaced by resistant ones. Consequently, someone in a hospital for wound treatment could find the wound infected with MRSA bacteria and no viable treatment available. The change in the bacteria is entirely consistent with what happens in evolution.

Contributing to the problem is what some have described as the irresponsible use of antibiotics in animal agriculture. Farming in general has become a highly sophisticated, very technical industry. The days of cattle wandering aimlessly through a pasture or a farmer tossing corn to his chickens that scratch around the barnyard are pretty much gone. Animals are now raised on what amounts to factory farms where they are often confined and crowded. Such conditions stress the animals, which may contribute to immune system dysfunction. Moreover, under those conditions the animals are exposed to urinary and fecal wastes, which are fertile breeding grounds for microorganisms, including pathogenic bacteria. Under such conditions, animals are routinely fed antibiotics perhaps prophylactically to prevent disease outbreak. In addition, some antibiotics actually contribute to animal growth, raising the price on the hoof and increasing profits. It has been estimated that 70% of antibiotic use in the United States is used in agriculture.

The eventual impact all of this will have on human health is difficult to predict, but most scholars agree that it will be bad. Some even predict that we will return to conditions similar to those that existed before antibiotics were discovered. It remains to be seen whether or not they are right, but it is known that diseases like pneumonia and tuberculosis that were once easily controlled by antibiotics are now often not. And because antibiotic development is expensive but the return is not nearly as lucrative as it is for drugs like antidepressants and statins, there is little incentive for pharmaceutical firms to engage in it. Consequently, it is probable that we will eventually see infection claiming a significant number of lives, if that is not happening now.

The issue of bacterial drug resistance in human health is an important one that goes much beyond its serving as another illustration of evolution. Fortunately it is being taken seriously, and the way antibiotics are used in treating people is changing, albeit slowly. Agricultural use may be another story. The subject is explored in a number of books. *The Antibiotic Paradox* by Stuart Levy (2002) and *The Killers Within* by Michael Shnayerson and Mark Plotkin (2002) are only two, but they are a start. Additionally, MRSA specifically is the subject of several others. The inescapable conclusion, however, is that in a changing environment, species of organisms sometimes change too. They do not do it deliberately; it is strictly a result of random genetic alteration in the face of a selective agent. The drug resistant bacteria and pesticide resistant arthropods demonstrate that it happens.

Selection for resistant bacteria gets more complicated when antibiotics are released into the environment. This happens both by run-off from animal feed lots and release from municipal sewage treatment facilities. In the latter case, antibiotic residues that are either poorly absorbed or incompletely metabolized are released from our bodies with urine and feces sometimes along with unused antibiotics that are deliberately discarded by flushing down a toilet. Some apparently pass through sewage treatment plants, along with other pharmaceuticals, cosmetics, and additional personal chemicals, and enter receiving bodies of water, such as large rivers that might be the drinking water sources of communities downstream. Conceivably, antibiotics in water resources could selectively eliminate susceptible waterborne pathogenic bacteria like those causing cholera and typhoid fever, thus selecting for resistant strains. In addition to that, some people have a tendency toward antibiotic allergies. Again conceivably, antibiotics in the water supply could sensitize someone to the point where treatment with antibiotics could trigger a reaction.

Agricultural loss of antibiotics can be a significant problem where confined animal feeding operations represent a major industry. Feedlot residues, animal wastes in essence, can leach into ground water and subsequently into wells or surface water even in the face of attempts at prevention. Antibiotic residues in the wastes travel along.

There is an avenue where the overuse of antibiotics can impact human health in addition to promoting the evolution of resistant bacteria and possibly causing allergies. Our bodies host massive populations of bacteria, some of which are beneficial, some benign, and some potentially harmful. The *Staphylococcus* on our skin or *Streptococcus* in our respiratory systems fall into the latter category as does a spore-forming organism named *Clostridium difficile*, usually referred to as *C. difficile*. This bug lives in our lower digestive system, where it is normally harmless because its population is held in check by competition with the more benign and beneficial organisms that also normally live there. However, treatment with antibiotics can kill off intestinal bacteria. Possibly because they form spores, dormant structures that allow organisms that produce them to tolerate stressful periods, *C. difficile* is somewhat naturally resistant to many antibiotics. It responds to the decrease in its competitors' populations with a rapid population increase of its own. This can lead to diarrhea or, more seriously, psuedomembranous colitis and the complications associated with it, including, possibly, death.

As mentioned earlier, similar stories can be told about DDT and insect resistance to it. Like the antibiotics, at first use DDT was extremely effective and many people owe their freedom from diseases if not their lives to the DDT that eliminated disease-carrying ticks and insects. To the malaria and typhus mentioned earlier, one can add yellow fever, kala-azar, West Nile virus, African trypanosomiasis, possibly plague, and more. In addition, DDT made agriculture far more productive by eliminating crop-destroying insects. Unfortunately, failure to foresee a dark cloud behind the pesticide silver lining led to the abuse and misuse of DDT. DDT was sold in hand-held aerosol cans for use around the home, where people sprayed for ants, spiders, wasps, and similar vermin. Municipalities sprayed DDT wholesale for mosquitoes and other insects regarded as pests. On farms, airplanes were used to spray DDT on crops. Pests like corn borers, coddling moths, and locusts were all but eliminated, and cosmetically perfect produce, once the exception, became the rule. Unfortunately, the relict pest populations that survived were able to reproduce and pass their resistance on to their progeny, particularly the rapidly reproducing mosquitoes. The result of the use of DDT was the selective elimination of the susceptible insects and the survival of the resistant.

There were additional side effects of DDT use. Among them, non-target, even beneficial insects were also affected, and many of them did not reproduce as quickly as mosquitoes; consequently, they did not develop a tolerance. Thus, pollinators like honey bees and bumble bees and predacious insects like preying mantises and dragonflies also fell victim to DDT. Not all died, of course, but those that survived carried pesticide residues in their tissues. Complicating this, DDT was resistant to bacterial decay, and so it persisted in the environment into which it was sprayed. DDT accumulated in soils and in the bodies of organisms it did not kill. As predatory animals ate DDT laden insects, they tended to concentrate the poison in their tissues. In birds this led to thinning of eggshells, which effectively brought about reproductive failure. In addition, DDT often failed to stay where it was applied, migrating instead to rivers and other water bodies and causing environmental disruption there.

DDT use in the United States was banned in 1970 owing to its potential harm to human health. There is disagreement over exactly what harm it does, although similar compounds, the so-called polychlorinated biphenyls, have been linked to reproductive anomalies and cancer. Many tropical countries still use DDT, particularly against malaria-carrying mosquitoes, but with less effectiveness than in the past. Indeed, it has been argued that the resurgence of malaria in tropical countries has resulted from the abandonment of DDT use, and there are those who would like to see its use resumed worldwide. But it is likely that mosquito resistance to DDT is contributing more to the resurgence of the disease than many realize, and reintroducing DDT would probably be less effective than its proponents think. Furthermore, the parasite that causes malaria is evolving its own resistance to antimalarial drugs, rendering them ineffective in some parts of the world.

In summary then, between World Wars I and II, the human race came into possession of major chemical weapons in the war against bugs: the kind that cause disease and the kind that crawl and can spread disease. The use of antibiotics and pesticides, respectively, has undoubtedly saved innumerable lives, increased levels of comfort and well being, promoted agricultural productivity, and supported a historically unprecedented growth of the human population on the planet. Simultaneously, these chemicals began a selective elimination of those organisms that were most susceptible to them, and in the process encouraged the developing and prospering of organisms that could withstand them. As a result, we now have populations of pathogenic microorganisms and disease-carrying arthropods that are insensitive to those chemicals, placing us in a situation somewhat similar to that which existed prior to having them. In other words, we are almost back where we started. What has happened in the advancement of these "superbugs" is entirely consistent with what happens when any species is confronted by some new selective agent. It either goes extinct or it evolves a tolerance. It does not have a choice between these options; either is the result of random biochemical events, just as evolution in the larger sense is the accumulation of random biochemical events. In this case, however, the change has occurred in less than a century, amounting to a rapid punctuation is what has undoubtedly been an equilibrium that has been in places as long as the species exited. The case of the superbugs demonstrates a biological flexibility that admittedly caught many scientists by surprise, but it is still more evidence that species of organism are capable of change over time, which is exactly what evolution amounts to.

References

Levy SB (2002) The antibiotic paradox. Da Capo Press, Cambridge

Plotkin MJ, Shnayerson M (2002) The killers within: the deadly rise of drug-resistant bacteria. Little, Brown Co, New York

Chapter 7
Biogeography

Chapter 11 of the Book of Genesis in the Old Testament[1] tells the story of the tower of Babel, where the descendents of Noah constructed a high tower, presumably to have a vantage point to see if another flood was coming. According to the chapter, all humans were of one people with one language, and God's reaction to the tower was to "... confound their language, that they may not understand one another's speech," (Verse 7) and "... scattered them abroad from thence upon the face of all the earth," (Verse 8). Regardless of how it was accomplished, there is no denying that humans succeeded quite well in spreading out upon the face of all the earth and in developing a multitude of languages.

As will be discussed in Chap. 14, humans originated in Africa and migrated to the rest of the world from there. Indeed, emigration occurred more than once, but the first emigration of modern humans from Africa is believed to have occurred around 60,000 years ago. At that time, the most recent ice age was perhaps halfway through its duration, and the planet was a much colder place than it is today. Thick sheets of ice covered much of the northern hemisphere. All of the water tied up in that ice had come from the oceans, of course, which means the sea level was substantially lower than it is now. With that being the case, shallow Babel Mandeb at the south end of the Red Sea that separates East Africa from the Arabian Peninsula and the Strait of Hormuz that separates the Arabian Peninsula from the South Asian mainland were most likely well above sea level. Travel from Africa to Asia by way of the Arabian Peninsula was entirely feasible.

As humans spread out over the earth, they encountered new environments with strange plants and animals. The initial contact, however, was gradual, and afforded those early humans time to adjust to the new environments and their inhabitants before totally breaking away from the old ones. Thus when humans first reached Australia, perhaps as much as 47,000 years ago, they had probably encountered comparable animals on the islands north of what is now the continent. At that time,

[1] *The Holy Bible*, King James version

B. Marcus, *Evolution That Anyone Can Understand*,
SpringerBriefs in Evolutionary Biology, DOI: 10.1007/978-1-4419-6126-6_7,

Australia was probably connected to the island of Papua New Guinea and some of the smaller nearby islands of the Malay Archipelago. Marsupials that recently have been restricted to Australia probably ranged through those islands as well, and the ancestors of the modern Australian Aborigines were most likely inured to them.

Such was undoubtedly not the case when recent Europeans first set foot on Australia, around 400 years ago. Those were Dutch sailors, and the first sighting of a kangaroo, an echidna, or a platypus must have left them with their mouths hanging agape in astonishment.

That Australia is home to a collection of bizarre animals is known well enough, although it is much more extensive than most people realize. In particular, Australia is known for its marsupial mammals, animals that give birth to young ones prematurely and then nourish them externally to the mother as their development continues. Nearby New Guinea supports its own population of marsupials, and opossums, another group, are found in the Americas, but Australia is considered to be the marsupials' continent. The baby marsupial must make its way from its mother's birth canal to a nipple, where it then completes its embryonic development. In the case of the kangaroo and some others, this takes place in a marsupium or pouch. Placental mammals, those that have long internal embryonic development, including ourselves, have arrived in Australia only recently, and with few exceptions, they were brought there, accidentally or deliberately, by Europeans.

One placental mammal, or more properly group of mammals, that got to Australia on its own was the bat. Bats are the most widely distributed of placental mammals for a reason that is obvious. They fly! This is not to suggest that bats deliberately migrated to Australia, although it is entirely possible that they spread down the Malay Archipelago from the Asian Mainland and eventually reached Australia as their populations expanded elsewhere in Asia. Another possibility is that they were blown to Australia by a storm, just as the ancestors of Darwin's finches are thought to have reached the Galapagos and the ancestors of the Nēnē got to Hawaii. Another non-human placental animal that is native to the island continent is the Australian wild dog, better known as the dingo, which most likely arrived in Australia from Southeast Asia. Whether it did so on its own by accidental rafting or with help from humans is not clear. Rodents easily could have reached Australia by rafting. Otherwise, the dominant mammals of Australia are indeed marsupials.

Another group of odd mammals that is found in Australia, along with a representative in New Guinea, is the monotremes. These include the duckbill platypus and a small variety of animals collectively referred to as echidnas or spiny anteaters. Monotremes have a most unmammalian characteristic: they lay eggs. Monotreme eggs are surrounded by a leathery shell like those of reptiles rather than by a calcium shell like those of birds. But the animals are clearly mammals. Their bodies are covered with hair, they are able to use metabolic heat to maintain a body temperature above that of the air around them, and the females nourish their young with milk. None of these qualities is characteristic of reptiles. Consequently, monotremes are often described as transitional animals; they represent a link between reptiles and mammals. This is not to say that they are *the* link between the two categories. In fact, they more likely are similar to an animal that evolved after reptiles and mammals diverged. But they are very much like an animal that would be transitional betweens reptiles and placental mammals, and it is probably correct to say that they represent a primitive mammal.

If all life was created in a single week and those animals in existence now are the descendents of the survivors of Noah's flood, animals of Australia pose a number of perplexing questions. For example, why did practically all living marsupials end up in Australia, how did they get there, and why did so few placental mammals get there when they managed to populate the rest of the earth? The same questions can be asked about the monotremes and the giant Gippsland earthworm of Australia, an earthworm that measures more than a meter in length. Why did so many unique animals gravitate to Australia and practically nowhere else, sometimes navigating highly hostile environments in the process? On a more grand scale, why does every continent appear to have its own unique zoology? Why did the animals leaving the Ark not simply wander off randomly? Cases in point, why are rodents so much more diverse in South America than they are in any other continent, why do they grow so large there but nowhere else, and why do old world and new world monkeys differ from one another more than do African and Asian monkeys? Why are there no tigers in Africa? Why are wolves found in both Eurasia and North America but mountain lions are found only in the latter? What about the giant tortoises and finches that are unique to each Galapagos island and the golden toads of Monte Verde, Costa Rica? The list goes on. Once again, science can be applied to answer these questions.

To begin with, scientists have determined that the world is far older than the 10,000 years or less suggested by the bible, as mentioned in Chap. 2. Using radiometric dating, a process where the abundances of radioactive materials and their breakdown products are compared and the known rate of the radioactive material's decay is applied, the age of the earth has been estimated to be more than 4.5 billion years. Life is thought to have existed on the planet for roughly three quarters of that time. Multicellular life more or less as we know it has been present for perhaps a billion years, but it was around 550 million years ago, the so-called Cambrian explosion, when the major animal groups that we are familiar with today first became abundant. This is according to the fossil record, which will be discussed in Chap. 10.

During the Cambrian period, the distribution of the earth's continents was nothing like it is today. Most of the land mass was south of the equator. Not that that made any difference in terms of the animals that existed then. They all lived in the seas. What is important about the continents is that they were not then nor are they now fixed in place. Rather, they tend to migrate over the surface of the planet, albeit very slowly.

The idea of wandering continents at first defies common sense, and for many years geologists, especially in the northern hemisphere, questioned the veracity of what is known as continental drift. However, as with many scientific ideas, determined research eventually yielded the truth about the concept, and it turns out that common sense was wrong.

The structure of the earth, in overly simple terms, can be compared to that of a peach. Just as a peach is enclosed by skin, the earth is covered by its skin, which is more properly identified as its crust. Beneath the crust is the earth's mantel, a layer of hot, molten rock that compares to the fleshy part of the peach. Finally, at the center of the earth is the core, which compares to the pit of the peach. Unlike the skin of the peach, however, the crust of the earth is broken into jigsaw puzzle-like components called tectonic plates. These essentially float on the molten mantel. Where the plates abut against one another, there is usually a lot of geological activity, such as earthquakes and volcanoes. This was demonstrated rather dramatically during the spring of 2010, when a volcano in Iceland, which sits on the juncture between two plates, erupted and spewed out an ash cloud that disrupted air travel in Europe. The entire island of Iceland is a geologically active area. Further south along the juncture, for example in the middle of the Atlantic Ocean, molten material from the mantel flows upward and adheres to the edges of the plates, forcing points on either side of the junction away from one another. Thus, the floor of the Atlantic Ocean is actually getting wider, and the landmasses on either side are being forced away from one another. One can almost see this by examining a world map and looking at the east coast of South America and the west Coast of Africa. The two continents appear as if they could fit together, and indeed, during the distant past, they did.

The opposite of these zones of sea-floor spreading are so-called zones of subduction, where plate edges are being pushed together, and one is essentially forced to slide under the other. This is happening in the Marianas Trench, the deepest point in the Pacific Ocean, where the Pacific plate is sliding under the Eurasian plate. At still other plate junctions, such as where the Pacific plate meets the North American Plate along the west coast of North America, the infamous San Andreas Fault, two plates move against one another in opposite directions.

During the last period of the Paleozoic era, the Permian, which ended around 250 million years ago, all of the earth's landmasses were clustered together into a single super continent called Pangaea that stretched essentially from pole to pole. During the Triassic Period, the first part of the Mesozoic era, which followed the Paleozoic, the northern and southern halves of Pangaea had pulled apart into two smaller continents: Gondwana in the south and Laurasia in the north. During the Triassic, according to the fossil record, both dinosaurs and mammals originated and had likely spread over as much available land as they could, although neither had yet reached its subsequent evolutionary diversity.

The southern super continent, Gondwana, began breaking apart around 150 million years ago, but Australia, Antarctica, and South America remained connected until around 60 or 65 million years ago. At that time, marsupials were quite diverse and widespread, presumable equally so in the southern and northern continents. This is consistent with the fossil record of South America being rich with marsupial species, one of which, the Virginia opossum, eventually migrated to North America where it still survives. No marsupials, however, are known to now exist in Europe, mainland Asia, or Africa.

Perhaps the first person to have noticed that the landmasses of the planet each appear to have their own characteristic collection of animals was Alfred Russel Wallace, Darwin's co-determiner of evolution. Wallace defined the concept of biogeography and divided the terrestrial world into eight realms based on their animal, principally mammalian, inhabitants. Thus, Australia and the nearby islands to its north became identified as the Australian, more recently Australasian, realm, which is typified by marsupials and monotremes. The Antarctic continent no longer has living marsupials, of course, and its characteristic inhabitants are the penguins. It is referred to as the Antarctic realm, and South America, which is the home of the world's greatest diversity of rodents, along with Central America north to southern Mexico is the Neotropical realm.

Sub-Saharan Africa with its great herds of grazing animals is identified as the Ethiopian or Afrotropical realm; Asia, from India east through Southeast Asia south of the Himalayas and the Malay archipelago is the Oriental or Indomaylayan realm; the remainder of Asia, North Africa, and Europe make up the Palearctic realm; North America north of southern Mexico is the Nearctic realm; and Pacific Oceana is the Oceanic realm.

Admittedly, drawing precise lines where one realm ends and the adjacent one begins is something of a challenge. In addition, there is considerable overlap in terms of the types and even species of animals found in many of them. While the Australasian, Antarctic, and Neotropical realms are reasonably distinct, the Afrotropical, Indomaylayan, and Palearctic realms share quite a few types of animals, and the Palearctic and Nearctic share many. The border between any two realms can be an impassible barrier, an ocean for example, or it may be a selective one, allowing some animals to pass but blocking others. In some instances, the boundary may actually be a passageway between two realms, and under the right circumstances, representatives of animal species migrate from one realm to another.

As just mentioned, the Nearctic and Palearctic realms share many similar animals. Both continents were part of the original super continent Laurasia, which began breaking up around 200 million years ago, well before the origin of modern mammals. Europe and North America both remained attached to Greenland until as recently as 65 million years ago, still before modern mammals had originated. However, that attachment may explain why colder water fish families like the pikes, perches, and salmons are common to both of the two continents.

While the Atlantic Ocean grew and separated Europe from North America, its level, along with those of the other oceans, fluctuated as glacial periods, ice ages, developed and waned. At some point, western North America and Eastern Eurasia migrated close enough to each other so that during glacial periods, when ocean levels fell enough, a land bridge between the two continents, specifically between what is now the Seward Peninsula of western Alaska and the eastern point of Siberia, existed.

The most recent ice age, the Wisconsin Glacial Period, lasted for 100,000 years. It sucked up enough seawater to expose a passageway, now known as the Bering Land Bridge, as much as 1,000 miles wide, that connected North America to Eurasia. Perhaps more importantly, it extended from east to west, allowing any organisms migrating across it to travel through a more or less uniform climate, in contrast to that which organisms traveling from north to south would encounter.

As a result, a substantial number of similar mammals now occur in the two major continents of the northern hemisphere. Examples include large animals like the North American bison (*Bison bison*) and the European wisent (*Bison bonasus*), two animals that differ slightly from one another but are obviously closely related. The American moose (*Alces alces*) actually inhabits Eurasia as well, where it is known as the elk. (The American elk or wapiti [*Cervus Canadensis*] is in reality a large deer, not a true elk.) Because it is found all through the subpolar Northern Hemisphere, *Alces alces* is described as a circumpolar species. Another species that ranges widely across the Northern Hemisphere is the brown bear *Ursus arctos*. In North America, the brown bear is represented by two sub species, the Kodiak bear and the grizzly bear. It is not truly circumpolar because it is not found in the eastern half of North America. However, the important point is that it is found on both northern continents, having found means of traveling from one to the other.

Obviously the faunas of North America and Eurasia are not identical. Not all immigrants to North America survived, the American lion and American elephant for example. However, lions and elephants no longer survive in the Palearctic realm either. Likewise, not every species of animal migrated, the American turkey for example, and at least two, the horse and camel, failed to survive in the land on which they originated, North America, but prospered where they emigrated. Even so, sub populations of the original emigrants often segregated themselves into isolated populations and diverged from their ancestors. In the case of the horse family for example, the European animal became the horse while animals that spread to Africa became donkeys and zebras.

The list of circumpolar animals continues. The gray wolf (*canus lupis*) probably came to North America from Eurasia via the Bering land bridge, but it appears to have become widespread only since the die-off of other predators that occurred at the close of the most recent ice age. It ranged broadly from then until the arrival of the Europeans during the last millennium, and since then it has been hunted severely. Another group of carnivores, the mustelids, which include otters, weasels, badgers, and the like are also well represented in both Eurasia and North America, as well as on other continents. Even squirrels have managed to spread over both of the northern continents. All in all, northern Europeans that explored Canada and the northeastern United States most likely found an environment and animals that were for the most part somewhat familiar.

The same probably cannot be said about the Spanish and Portuguese explorers who ventured into Central and South America and perhaps even those exploring the Caribbean Islands and the southeastern United States. To be sure, the large animals they encountered in what is now Florida and the Gulf Coast, with the exception of the American alligator, were similar enough to those wandering around the rest of North America, but those of the American tropics, the Neotropical Realm of Wallace, were unique.

Recall that prior to around 250 million years ago, all landmasses on Earth were arranged in the single supercontinent Pangea, and Pangea broke up over a long period of time. By the time the mammals originated, perhaps 200 million years ago, the super continent was separating into the northern and southern continents,

Laurasia and Gondwana respectively as mentioned earlier. Once mammals originated, they were able to spread across the Earth, using land bridges where available and rafting to get to continents that had broken off from the fragmenting super continents. Mammals most likely remained small and inconspicuous, as the world was dominated by dinosaurs at that time. Marsupials split off from the main line of placental mammals more than 90 million years ago. Presumably they originated somewhere in Laurasia but managed to find their way to South America, possibly by way of a land bridge that stretched along the Caribbean Islands between what is now Florida and Venezuela. South America remained connected to Australia by way of Antarctica long enough for marsupials to have migrated there as well.

Most of the time since the break up of Gondwana, South America has been isolated. Consequently, once mammals got there, they had time and opportunity to uniquely diversify, and South America shows what may be the most unique collection of mammals on the planet, in terms of both the animals living there during the more distant geologic past and those living there now and during the recent geologic past. As noted earlier, marsupials survived in South America for much of the Cenozoic Era, and they diversified greatly and left a rich fossil record of their existence. Unique placental mammals evolved there as well, including the edentates, which contains the armadillo, sloth, and anteater. Monkeys and rodents arrived in South America around 40 million years ago, presumably by rafting and/or island hopping from Africa (Poux et al. 2006). The two continents were, of course, closer to one another then than they are now. Neotropical monkeys evolved along a pathway that differed from that of their ancestors in Africa. New world monkeys as a group became totally arboreal, unlike African monkeys that include a number of ground-dwelling varieties, and they developed prehensile tails. Moreover, while old world monkeys developed full color vision, new world monkeys for the most part did not. But they remained less diverse than their African relatives. On the other hand, rodents blossomed in South America, evolving into many singular species, including the capybara, which can weigh over 100 pounds. In addition, birds became quite diverse there as well.

The end of South America's isolation was no doubt a slow process that occurred with the formation of the Isthmus of Panama and the rest of Central America, perhaps between Guatemala and Columbia. Central America sits near the junction of two tectonic plates: the Caribbean and the Cocos. Most of Central America is actually on the Caribbean plate, with the subduction zone of the Cocos plate lying just off the Pacific coast. As with other tectonic junctions, Central America is very active geologically, with earthquakes and volcanic eruptions being quite common, much like the geologically active regions, say, around the island of Hawaii. Indeed, it is likely that the beginning of the formation of the Central American isthmus involved first the formation of an archipelago of volcanic islands. Over time, probably millions of years, the chain of islands coalesced into the narrow strip of land that now exists, being periodically submerged and exposed as ocean levels rose and fell with glaciations waxing and waning. The Current linkage between North and South America is largely believed to be around three million years old, and animals have been migrating between the two continents continually. Not all migrants succeeded, of course, and some animals never managed to make the crossing. Beavers, mountain sheep, and North American antelopes, for example, never made it to South America, and South American animals like the capybara never came north. If they did, they failed to survive. Thus, rather than acting as a broad highway allowing an essentially unimpeded exchange in both directions like the Bering land bridge, the isthmus of Central America instead acted as what ecologists call a filter route, allowing some animals to pass but holding back others (Feldhamer et al. 2007). It is not only the narrowness of the Central American isthmus that restricted animal travel between North and South America; it is also a matter of climate. The tropical heat of Central America most likely acted as a barrier to some animals going in each direction, as did an absence of specific foods. Animals that require a particular food that does not grow in a rain forest obviously would not survive there.

Of the animals that did manage to cross the Central American isthmus, those going south appear to have had more of an impact than those going north. Of the marsupials, coming north, the Virginia opossum is the only one that occupies much of a range north of Central America today. Armadillos live along the gulf coast, and New World monkeys, sloths, and anteaters live in the rain forests of Central America. During the Pleistocene, a number of other South American animals migrated north, including two quite large relatives of the modern anteaters, armadillos, and sloths: the giant ground sloth and the glyptodont. The South American version of the ground sloth was the weight of an elephant and, when standing on its hind legs, was able to reach as much as 6 m, almost 20 ft, off the ground, where it fed on leaves of trees. The North American version was not much smaller. The animal was common in the southern United States and may have ranged as far north as Canada. It went extinct around 10,000 years ago from the mainland and perhaps as recently as 5,000 years ago from some Caribbean Islands. It is believed that the human arrival in the New World dealt these animals their death knell.

If you can picture something that looked like a cross between an armadillo and a turtle that was about the size of a Volkswagen beetle with a head resembling a hornless bison's, you have a mental picture of a glyptodont. It too may have suffered its extinction at the hands of the humans who eventually migrated to the Americas.

Animals migrating from North America southward fared better. Ancestors of the South American camelids, without goats and sheep to compete with, were able to adapt to niches in the mountains, as were deer. Mustelids also diversified well in South America. Types living there now include relatives of raccoons, weasels, skunks, and otters including a species that reaches a length of 1.5 m (5 ft). Moreover, the large, predatory placental mammals that crossed the land bridge into South America found fierce competition among the native predators, but they managed to overcome those animals and drive them to extinction. Descendents of the pioneering carnivores survive today as the spectacled bear, maned wolf, and ten species of cats including the jaguar.

To reiterate, South America was an isolated continent for almost all of the Cenozoic, enough time for a completely unique zoology to have evolved there.

In addition to the animals already mentioned, there was an extensive radiation of marsupials, many of which evolved into fierce predators, including one that closely resembled North America's saber-toothed cat. In addition, South America also saw the evolution of a large, predatory bird, the so-called terror bird.

The South American saber-toothed marsupial, known by the genus name *Thylacosmilus*, was a predator that occupied the top of the food chain, just as large predators do today. Fossil remains suggest that it attacked by ambush rather than running down its prey, but its long canine teeth were most likely more than capable of quickly dispatching the prey when caught. It likely ruled much of its realm through the Pleistocine and early Miocine until the land bridge between North and South America formed. Once the more competitive North American predators were able to invade South America, *Thylacosmilus* became extinct. The animal may actually not have been a marsupial but rather a member of an order of mammals closely related to the marsupials. However, it was still a unique genus having appeared on no other continent. That it evolved to resemble the saber-toothed and scimitar cats of North America, true felids like modern lions, is the result of an evolutionary phenomenon known as convergence. This is an important concept that will be discussed in detail in Chap. 12.

The other apex predator of South America was the terror bird of the family Phorusrhacidae, a large, flightless bird that makes today's most fierce eagle look something like a parakeet. Actually, there were several species of this animal, which possibly originated in Antarctica but radiated into the several species in South America, one of which reached a height of 3 m (nearly 10 ft). Fossils suggest that the bird had a hooked beak, much like those of modern raptors, and it probably chased down its prey. Once the Isthmus of Panama formed, terror birds managed to spread into North America as far as Texas and Florida, but they were unable to survive there and were subsequently eliminated from South America as well. It is likely that they were unable to compete with mammalian hunters they encountered north of Panama, which eventually made their way south.

An alien arriving on Earth today would see a distribution of animals and plants quite different from that which would have naturally occurred on the planet had humans never appeared. Human beings have successfully spread their domestic animals onto every continent, and they have redistributed many wild animals as well. Consequently, one can now catch Western American rainbow trout in South America and New Zealand, hunt Chinese pheasant in South Dakota, and shoot European wild boars in Australia and the United States. Some organisms have made their way around the world as camp followers of humans or hitchhikers on international shipping. Thus the Australasian brown tree snake, the Norway rat, and the zebra mussel have made their way into new habitats around the world. Often, animals and plants in new habitats, whether introduced accidentally or deliberately, can wreak havoc, either by filling a vacant niche or out competing native species for their niches. The European rabbit, for example, has caused serious soil erosion in Australia, where, with no real natural enemies, it has reproduced almost without limit and has eaten away the vegetation that has normally held the soil in place. Likewise, Eurasian milfoil, since coming to North

America, has grown profusely and is now a problem in many lakes and other waterways. And Asian carp have made their way up the Mississippi River and now threaten to disrupt the ecology of Lake Michigan.

There is an idea among many people that the natural world exists in some kind of peaceful equilibrium some refer to as the balance of nature. The idea goes on to explain that every plant and animal serves some kind of purpose in terms of maintaining the equilibrium. While the thought is attractive and from superficial observation would appear to be correct, the real world is somewhat more complex. Natural systems, ecosystems as biologists know them, tend to be in a constant state of flux. An abandoned farmland in the northeastern United States, for example, over a course of years will become a meadow, a field of brush and shrubs, a pine woodlot, and eventually a broadleaf woodlot. The broadleaf woodlot may persist for many years, as long as the environment remains stable. But an outbreak of leaf-eating caterpillars, a fire, or a flood can seriously disrupt or even destroy even the most stable of systems. Over a long enough period of time, such disruptions are inevitable. However, if the ecosystem is not totally destroyed, there is a reasonable chance that it will recover and return to something that is, if not identical to the previous ecosystem, similar to it.

When a new plant or animal is introduced into an ecosystem, however, it can throw whatever stability exists into disorder. Thus, for example, when European chestnut trees and lumber were imported into the United States a bit over a century ago, a fungus the European species carried attacked American chestnut trees and virtually wiped them out over the next 50 years. The long association between the European trees and the fungus had allowed the two to evolve a mutual accommodation, which could conceivable have happened with the American trees and the fungus. However, when it appeared that the trees were blighted, timber interests harvested the trees extensively, essentially accomplishing the extinction that the fungus was in the process of doing.

The so-called balance of nature mentioned above is actually a complex of interaction and accommodations that will be described somewhat in the next chapter. These have developed over long periods of time and are as much a product of evolution as any individual organism happens to be. Thus, each of the continental realms that Alfred Russel Wallace described was the product of millions of years of isolation, gene mutation, species interaction, and even natural catastrophe that shaped the biological world until human activity reshaped it into what we have today.

References

Feldhamer GA, Drickamer LC, Vessey SH, Krajewski C (2007) Mammalogy: adaptation, diversity, ecology. The Johns Hopkins University Press, Baltimore

Poux C, Chevret P, Huchon D, de Jong WW, Douzery EJP (2006) Arrival and diversification of caviomorph rodents and platyrrhine primates in South America. Syst Biol 55(2):228–244

Chapter 8
Up a Blind Alley

A number of years ago I attended a talk by Professor Bruce Sundrud of Pennsylvania's Harrisburg Area Community College. In the talk, Professor Sundrud described a relationship between plants of the genus *Yucca* and a moth of the genus *Tegeticula.* After mating, the female moth, which Professor Sundrud called the "mother moth," gathers pollen from the flower of one *Yucca* plant and carries it to another. There she packs it onto the pistil, the pollen-receiving organ, of the recipient flower after first having laid an egg on it. The moth larvae, once it hatches, will feed on the pollen, but in the meantime, the "mother moth" has succeeded in pollinating the *Yucca* plant. After describing the behavior of the moth, Professor Sundrud looked around and said dryly, "That's a pistil-packin' momma."

While the audience got a laugh out of that, what Professor Sundrud had described was anything but comical from the points of view of the two organisms involved in a relationship of mutual dependence. If anything happens to eliminate one of the species, the other is doomed. It turns out that the moth will gather pollen in which to lay its egg from no other plant, and no other insect transfers pollen from one *Yucca* flower to another. The two organisms are completely dependent upon one another in order to complete their respective life cycles.

Nature is replete with examples of organisms that are solely dependent upon a single other species in order to complete their life cycles. Biologists describe such relationships as symbioses (singular: symbiosis), and they typically take one of three different arrangements. In the case of mutualistic symbiosis or mutualism, each organism benefits from the association with the other. One example of that would be the *Yucca* and yucca moth described above. Another would be the clown fish and sea anemone. The clownfish is left unharmed by the stinging tentacles of the anemone, among which it shelters and is kept safe from predators. In return, the clownfish may lure predacious fish to the anemone, which then stings and eats them. Clownfish also eat organic debris from food that is left over from the anemone's meal, thus keeping the environment around the anemone from fouling. It may also

B. Marcus, *Evolution That Anyone Can Understand*,
SpringerBriefs in Evolutionary Biology, DOI: 10.1007/978-1-4419-6126-6_8,

chase away butterfly fish, which will eat an anemone's tentacles. Sometimes a mutualistic relationship will be very general, as in the case of honeybees that pollinate a variety of flowers. In others it may be highly species specific. Plants of the family Fabaceae, commonly known as legumes, form a relationship with soil bacteria, collectively known as rhizobia, which have the capability of converting atmospheric nitrogen into a form the plants can use. Plants cannot obtain nitrogen from the air themselves and must rely on soil or symbiotic bacteria. The legumes and rhizobia secrete chemicals that attract one another, and the rhizobia make their way to the legume roots, which they infect. The legume forms nodules around the infecting bacteria and provide nutrients for which the rhizobia exchange nitrogen they have captured from the air. Each species of rhizobia has a species of legume that it infects.

The greater the diversity of organisms in a given environment, the more examples of mutualistic symbiosis one can find. Among the most diverse environments are the tropical rain forests, and within those one finds many examples of plants with dedicated pollinators, as specific as the *Yucca* and yucca moth, and plants with specific root mutualists. In addition, one can also find examples of animals that have undergone a modification of their behavior in order to carry on a relationship with a specific plant. Premiere among these would be the ants.

In the Amazon forests of Southeastern Peru, one can find ants that farm. Among them are arboreal species that seek out the seeds of specific epiphytes and literally plant them in what is described as ant gardens, actually the ant colonies, high in the trees. Epiphytes are plants that grow supported on other plants. They do not have contact with the ground; consequently, their seeds have the challenge of finding a suitable place on a tree limb to geminate. Most accomplish that by producing many small seeds that can be easily blown around by the wind. Alternatively, some epiphytes produce seed-filled fruits that are eaten by animals like bats and monkeys. When the animal defecates, the seeds are carried by the feces, which may land on a branch. In both cases, many seeds do not find a suitable place to germinate, and the energy that went into producing them is wasted. In the case of epiphytes planted in ant gardens, however, seeds are carried by ants from the parent plant or from the jungle floor to the ant colonies in the trees, where they are deliberately planted. The roots of the growing plant reinforce the walls of the colonies, holding them together, and remove excess moisture from the colony. In return, the plants enjoy a high rate of germination (Youngsteadt et al. 2009). There are a number of ant and plant species that have this kind of relationship, but each ant species will farm only a specific species of plant, which suggests that the phenomenon originated several times each independently of the others.

The arboreal ants are not the only farmers among the ants. A number of species grow fungal gardens in the soil, including a fascinating variety known as leaf-cutters. These ants are a common sight in tropical American forests and savannas, where they form well-worn trails on which the ants can be seen carrying leaf fragments that they have clipped off of some tree. The ants carry the leaf clippings back to their underground nests, where they use them as a substrate on which to cultivate a specific species of fungus. The ants tend the fungus gardens, removing any other species that may try to grow there. Although the fungus is grown to be eaten by the ants, its existence depends entirely upon the ants. The species cultivated by the ants grow nowhere other than the ant colonies and unless tended by the ants are quickly overwhelmed by other fungus species. For their part, the ants eat nothing other than the fungus they grow, with the possible exception of sap from the leaves they cut. The ant and fungi species are mutually dependent upon one another. Neither could exist without the other.

Still other species of ants enter into mutualistic relationships with acacia bushes. The bushes secrete nectar that attracts the ants, and they provide hollow thorns in which the ants can live. In return, the ants protect the bushes by attacking any animal that tries to eat the bush.

Mutualistic symbioses are very common. The more one looks, the more one finds, particularly in areas of high biodiversity. Indeed, if one starts examining the variety of microbes that have mutualistic relationships with larger organisms, such as the wood-digesting organisms in a termite's gut, the number becomes immense. The participants in a mutualistically symbiotic relationship are not always completely dependent upon their partner, and the extinction of one will not necessarily bring about the extinction of the other. Conceivably, rhizobia bacteria can live independently in soil, and legumes, provided the soil is sufficiently rich in nitrogen, can exist without symbiotic rhizobia bacteria to fix nitrogen for them. But both organisms survive more successfully when they live together. However, as with the *Yucca* and its pollinating moth and the leafcutter ants and their fungus, mutualistic symbiosis can become mutual dependency. Each partner sometimes surrenders its independence from the other in order to gain a benefit. Therefore, the evolution of this relationship can figuratively take the participating partners up a blind alley; there is no way out. At this moment, it would be helpful to cite an example of a coextinction of a pair of mutualists that has occurred to illustrate the point. One may actually be in the process of happening at the time this is being written.

The Hawaiian woody plant *Hibiscadelphus* is usually pollinated by endemic birds known as honeycreepers. The closed flowers allow honeycreepers to get their long, curved bill inside, but other potential pollinators are unable to. Since the arrival of humans on Hawaii, honeycreepers have been in decline, and *Hibiscadelphus* has been declining as well. There is no guarantee that the decline of one has caused the decline of the other, but the coincidence is more than striking.

In some cases, mutualism can be so thorough that the two organisms involved appear to be only one. The classic example of this, the one that is cited in most textbooks, is the lichen, a small, primitive plant-like organism that is actually a composite of an alga and a fungus. Lichens often grow where true plants will not, such as on bare rock, and they are often the first organisms to colonize an area that has been sterilized. Neither the alga nor the fungus would be able to survive in such environments on its own, but together in the lichen they appear to do quite well.

Even more bizarre than a pair of organisms function as a single one is a colony. That is the case with the Portuguese man-of-war, a jellyfish-like creature that one finds in tropical seas. This organism, for want of a better word, consists of a number of individuals living together as one, with each one taking on specific roles and acting almost as organs. One individual is essentially a gas bladder, called the sail, which floats on the water's surface. The remainder of the colony is suspended from the sail, essentially as tentacles but with a variety of tasks. Some individuals are responsible for obtaining food with stinging cells that contain venom that paralyzes anything that happens to float or swim up against them. Others transport the prey to digestive structures that break down the prey tissues and distribute the nutrients obtained from the tissues to other members of the colony. Still other individuals are responsible for reproduction, producing sperm on some colonies and eggs on others. The sperm and eggs are released into the water where they meet and fuse, and a new individual develops. This individual undergoes budding, a form of reproduction that does not require two partners. The organisms that result from the budding form the colony, with each member of the colony taking on some specific task. It is almost as if the colony is trying to be a single organism but has not yet reached the level of organization of one.

University of Massachusetts biologist Lynn Margulis has written extensively on the subject of symbiosis and its role in evolution. Her 1999 book *Symbiotic Planet* (Margulis 1999) is recommended to those who would like to read more on the subject.

A second case of symbiosis is commensalism, where two organisms have a relationship in which one benefits and the other remains unaffected. Cattle egrets, the white birds that one sometimes see at the feet of cattle out in pastures, benefit from their association with cattle by eating insects the cattle stir up as they wander around. The cattle, however, neither gain nor lose from the birds. It would make no difference to them whether or not the egrets were present. Other examples include the Spanish moss and old man's beard one sees on trees in the Southeastern United States. These plants are epiphytes just as are the ones described earlier in the chapter in ant gardens. They use trees for support and to hold them high enough in the air where they can get light and not have to worry as much as ground plants about being shaded, but they obtain moisture and nutrients from the air, and not from the tree that supports them. Therefore, the tree remains unaffected.

Commensalistic relationships can be complicated. In tropical America, colonies of a variety of ants known as army ants periodically swarm through the rainforest devouring every small animal they can. Stories about their eating large animals like cows and people are gross exaggerations, but they do bite painfully, and large animals learn to stay out of their way. If a swarm of army ants moves into somebody's home, for example, they usually occupy only one or two rooms, and for two days or however long until the swarm leaves, those rooms are left alone. Once the ants are gone, the homeowner will find that the rooms have been sterilized of all pests that might have been living there, including cockroaches, mice, spiders, and even scorpions. All along the forest floor, animals scatter when army ants swarm, which presents opportunities to a number of bird species that follow the swarms and feed on the insects and some other small animals that are disrupted by them. Collectively known as antbirds, some species are highly dependent and the swarms as their principal source of food. Identified as obligate ant-followers, these species often keep an eye on the ant colonies and call when the colonies start to forage. Other birds, opportunistic ant-followers, are not dependent upon army ant foraging but take advantage of it when it occurs. These rely on the calls of the obligate ant-followers to alert them (Chaves-Campos 2003). None of the birds in any way impact the ants. However, both obligate and opportunistic ant-followers benefit from the ants, thus making them commensalistic with the army ants. To complicate the relationship even more, the opportunistic ant-followers are commensalistic with the obligate ant-followers in that the calls of obligate birds benefit the opportunistic ones, but the obligate birds are not impacted in return.

Pure dependency of a commensal on its independent partner is not as easy to demonstrate as it can be with mutualistic partners. Consequently, it may not be the blind alley that mutualism can be. However, should army ants ever become extinct, obligate ant-followers would be severely stressed. Indeed, it is likely that many individual birds would starve to death and some species would become extinct as well.

There are other commensalistic relationships where the benefiting partner would also be severely stressed, possibly to the point of extinction, with the loss of the unaffected partner. Australian koalas, for example, live only among eucalyptus trees. In contrast, eucalyptus has been successfully planted outside of Australia in the absence of koalas. Koalas' principal food is eucalyptus leaves and bark. Although they are known to eat other kinds of leaves, their being found only among eucalyptus suggests a strong dependence upon them. The relationship benefits the koala not only in that it has a source of food in the eucalyptus, but also in that it has little competition from other leaf-eating animals. Eucalyptus leaves are filled with a variety of chemical compounds that are poisonous to most animals. At first it would appear that the eucalyptus would be harmed by its relationship with koalas, but koalas tend to be territorial. Each koala has a home range, and others usually respect the range and avoid the trees within it. If anything were to happen to the eucalyptus trees, it is likely that koalas would die out. In contrast, the success of eucalyptus plantations outside of Australia, where koalas are absent, demonstrates that the extinction of koalas would not hurt the trees.

The third recognized type of symbiosis is parasitism, where two organisms live together for the benefit of one and to the detriment of the other. Parasites may live within their hosts as endoparasites or outside of them as ectoparasites. Some ectoparasites live on their hosts; others visit when it is time for a meal or when they wish to lay eggs. In at least one case, parasitism has evolved to the point of insuring reproduction. In a family of deepwater anglerfish, the small males permanently attach themselves to much larger females and undergo what amounts to developmental degeneration. Their eyes and internal organs fall into disuse, and their skin and circulatory systems fuse with those of their hostess. As the female continues to grow, the male essentially deteriorates to little more than an animated gonad. Undoubtedly, some readers will find that metaphorical to humans, but it is a successful lifestyle that guarantees reproduction of the anglerfish and illustrates that parasitic relationships can be complex and even bizarre. Some parasites need series of hosts in order to complete their life cycles; others need them for only part of their lives.

In general, people are disdainful of parasites, considering them to be somehow arrested in their development or evolution. In reality, the apparent arrested or degenerated state of development, as shown for example by the male anglerfish described in the preceding paragraph, actually represents a stage, often the final one, of the developmental sequence of the parasite. An organism that lives within its food supply has no need for the organs necessary to find and process that food. It is equally incorrect to think of parasites as evolutionarily arrested too. Parasites typically coevolve along with their hosts. It is for that reason that some parasites are so specialized that they will infect only a single species of host organism.

Parasites are often associated with illness and rightly so. Infectious diseases that have plagued humankind for millennia, including malaria, smallpox, and cholera, and newer ones like AIDS and Ebola, are caused by parasites. While many animals and plants can and do support a population of parasites with no obvious ill effects, if the parasites become too numerous, the host's health can deteriorate, even to the point of death. But killing its host does not serve the parasite's interest, especially endoparasites, because the death of the host often spells the death of the parasites it carries as well. Parasites want to keep their host alive at least until they have had the chance to reproduce. Moreover, most parasites are more than simply infectious organisms.

Parasitism takes on a variety of forms. Among insect parasites, for example, one finds wasps that could be described almost as predacious rather than parasitic; thus it is often referred to as a parasitoid rather than a true parasite. Each wasp species has a specific prey that it uses, and when a female wasp is ready to lay her eggs, she will search out her prey species, and sting it. Prey may be anything from an aphid, in the case of really small wasps, to caterpillars, large spiders or cockroaches. Once the prey is paralyzed by the sting, the wasp may lay one or several eggs, depending upon the species, and then either drag the prey off to her nest or allow it to recover. In either case, the eggs will hatch, and the larvae will start eating the host from the inside out. Usually the host lives long enough for the larva to complete its development. When the larva emerges from the host, the host dies.

Parasitic flies show a greater variety of behaviors toward their hosts. Some, like some species of the fly family Phoridae, are parasitoids of ants. One species in particular holds promise for controlling fire ants in the southern United States. The female fly lays an egg on a fire ant. Once the larva hatches, it makes its way to the interior of the ant's head and eats it. When the larva is finished, it pupates and emerges from the ants head. Needless to say, the ant is killed by the fly. Other flies act more like true parasites, and some infect mammals, including humans. The bot fly is just one example.

Bot flies may lay their eggs directly on their hosts, on biting flies like mosquitoes that prey on their hosts, or as in the case of the species that infects rabbits, at the entry into the hosts' burrow. In any case, the egg or larva once the egg hatches, finds its way to its mammal host and develops under the skin until it is full grown. It then usually drops off and pupates in the soil. In Central America, bot flies may be serious pests of cattle causing ranchers economic problems. Other flies that parasitize mammals, like the screwworm fly, have similar life cycles. Still other flies have similar relationships with still other vertebrates and invertebrates, including other insects.

Still other arthropods such as true bugs, lice, fleas, and ticks parasitize warm-blooded animals. Often referred to as ectoparasites because they live on, not in, their hosts, these do not lay eggs on their host; instead they suck blood, as anyone who has ever encountered them can attest. Most of these animals are more annoying than anything else, but they are capable of doing damage in terms of introducing other parasites into the host, either directly or indirectly. Many of them carry causative agents for infectious diseases that have plagued mankind for centuries, if not millennia. In some cases, such as Yellow fever, West Nile, and Eastern Equine Encephalitis, the causative agent is a virus, and the mosquito that transmits each disease is a simple carrier, referred to as a vector. Sometimes, as in the case of yellow fever, the virus relies on only a single species of mosquito as its vector. Other viruses may use several vector species within a given genus, and some may use a variety of vector species.

As do viruses, other disease-causing organisms may use biting flies or other blood-sucking arthropods as vectors. In some cases, the life cycle requires the vector to serve as an alternative host as well as a transport mechanism. In the case of malaria, for example, the parasite, a protozoan, requires a mosquito of the genus *Anopheles* in which it must undergo part of its development. Another protozoan causes Chagas disease, a form of endocarditis common in tropical America that may have affected Charles Darwin. The vector is a true bug that typically bites a sleeping victim, not always human, and simultaneously defecates. When the victim scratches at the itchy bite, he often contaminates the would with bug feces, and, if a Chagas disease protozoan is present in the feces, accomplishes infection.

In cases where the alternative host is not a biting arthropod, the principal host, from our perspective, need only be nearby. An illustration of this would be schistosomiasis, also known as bilharzia, where the parasite is a flatworm known as a fluke. It spends part of its life cycle in a vertebrate host like a human and part in a snail. Once the parasites in the snail mature, they leave their host, often killing it in the process, and seek out their alternative host. Sometimes they find the wrong one. Along the east coast of the United States, a form of dermatitis known as swimmers itch occurs when a particular species of these parasites, after leaving the snail find a human host instead of their natural host, some kind of marine bird. The worms are not able to complete their life cycle within a human, and they eventually die, usually causing no harm other than the irritating itch. This is not to say that all parasites are harmless to an accidental host. Trichinosis, a roundworm infection that comes from eating undercooked pork, is normally a parasite of rats. It can get into pigs if a pig happens to eat an infected rat. Pigs are omnivores; they will eat meat if they can get it. If a human eats undercooked pork infected with trichinosis worms, he can become very ill.

This discussion on parasites barely scratches the surface of a strange and fascinating world; one that is often so bizarre that it is difficult to believe. Indeed, some parasites have life cycles that defy imagination. One of the best descriptions I have seen of parasitism is *Parasite Rex* by science writer Carl Zimmer (2000). Again, the reader is encouraged to pursue this topic, and Zimmer's book is a good place to start.

While it is difficult to imagine how many of the symbioses described here may have come about, it does not change the reality that these relationships exist and that the diversity of life is far richer for them. Indeed, it is argued by some scientists that the diversity of life may be the result of ancient symbioses. If one looks inside the cell of a human or any other multicellular animal or plant, one sees a variety of structures that carry on specific jobs, much like the different members of a Portuguese man-of-war colony carry on specific jobs. One set of structures in particular, the mitochondria, merit a closer look. These sub-cellular structures, organelles, are the sites where cells take apart food molecules to release energy. In our cells and those of virtually all oxygen requiring organisms, it is here that oxygen combines with sugar molecules to form carbon dioxide. What is most fascinating about mitochondria is that they contain their own DNA, which is different from that found in the nucleus of the cell. Moreover, the mitochondria reproduce independently of the rest of the cell. When a cell divides into two, the mitochondria are simply randomly distributed between the two new cells. The same is true of chloroplasts, the green organelles in plant cells where photosynthesis occurs. You may recall that these were described in Chap. 4 in the discussion on fossil genes. Theory has it that both mitochondria and chloroplasts are the descendants of once free-living organisms that entered into symbiotic relationships with primitive cells and eventually lost their autonomy, much like the male anglerfish described earlier loses its autonomy when it becomes parasitic on the female. To be sure, there is at the moment no way of demonstrating how this may have happened, and there are some problems with it, but like any good theory, it is supported by evidence (Dyall et al. 2004). One example would be the single-celled organism *Paramecium bursaria* that when examined under the microscope appears to have chloroplasts. On closer examination, however, those chloroplasts turn out to be cells of endosymbiotic algae. The same is true of choral and other marine organisms.

Many will read this and find the idea that the relationships I described could have evolved randomly to be far-fetched. To them it is much more reasonable to simply attribute all of these phenomena to a creator. Others would ask why a creator would bother to form so many odd and convoluted relationships. Why create a Portuguese man-of-war as a colony, and not a single organism? Why put algae in *Paramecium bursaria* but in no other species of *Paramecium*, and why bother with algae? Why not simply use chloroplasts? That is the function they seem to serve. Indeed, why make a male anglerfish parasitic on the female? Why make parasites at all, and why make the life cycles of some so complex? And why create situations where an organism is vulnerable to extinction from the loss of its symbiont? From a scientific perspective, symbiotic relationships work, at least for one member. If they work for both, so much the better, but even a parasitic relationship works if the host is not destroyed before the parasite can complete the component of its life cycle for which the host is required. Any organism that gets involved in a relationship where it is obligatively dependent on another organism, as are many that are described in this chapter, are vulnerable should the organism upon which they depend cease to exist. But it is common because it works. In nature, things that work persist. Simply put, that is natural selection.

References

Chaves-Campos J (2003) Localization of army-ant swarms by ant-following birds on the Caribbean slope of Costa Rica: following the vocalization of antbirds to find the swarms. Ornitol Neot 14(3):289-294

Dyall SD, Brown MT, Johnson PJ (2004) Ancient invasions: from endosymbionts to organelles. Science 304(9):253-257

Margulis L (1999) Symbiotic planet: a new look at evolution. Basic Books, New York

Youngsteadt E, Baca JA, Osborne J, Schal C (2009) Species-specific seed dispersal in an obligate ant-plant mutualism. PLoS ONE 4(2):e4335. doi: 10.1371/journal.pone.0004335. http://www.plosone.org/article/info%3Adoi%2F10.1371%2Fjournal.pone.0004335. Accessed 21 Feb 2011

Zimmer C (2000) Parasite Rex: inside the Bizarre world of nature's most dangerous creatures. The Free Press, New York

Chapter 9
What is a Species?

A number of years ago, I had a student who was convinced that she had a pet that was half cat and half rabbit. She was certain that her pet had a parent of each species, and when I could not convince her that was impossible, I challenged her to bring the animal in so I could have a look at it. As it turned out, the animal was a variety of stub-tailed cat known as a Manx, which we will talk about in the following chapter. I offered to open it up and show her that its internal anatomy was all cat, but she declined, ostensibly, at any rate, accepting my argument.

Rabbits do not mate with cats, raccoons do not mate with woodchucks, horses do not mate with cows, and let us hope people stick to their own species as well. There are many reasons for this, among which we can list species-specific mating behaviors, biochemical or behavioral signals within a species, and the general impossibility of sperm of one species and egg of another from fertilizing one another. But it is at the species level where much of what we consider to be evolution occurs. Indeed, when members of a population have evolved to the point that they no longer can or will mate with members of similar, related populations, we often say that speciation has occurred.

Species represent one level of classification of living things with which biologists deal. Most people have heard of what many believe to be the broadest level of classification, the animal and plant kingdoms, and consider them to be all-inclusive in terms of the organisms they contain. In reality, biologists have broader classifications, but kingdom will work as a starting point for our purposes.

Kingdoms are divided into phyla (singular, phylum), which are more restrictive. For example, the phylum Arthropoda includes all animals with paired, jointed legs and a hard outer body covering. Both spiders and insects are members of this phylum. However, phyla are divided into still more restrictive categories called classes. Many people think spiders are insects, but they actually fall into a different class. The insect class includes those Arthropoda with three obvious body regions, three pairs of jointed legs attached to the middle region and, at some point in their lives, wings. In contrast, spiders, belonging to the class Arachnida, have two body

B. Marcus, *Evolution That Anyone Can Understand*,
SpringerBriefs in Evolutionary Biology, DOI: 10.1007/978-1-4419-6126-6_9,

regions, eight legs attached to the forward region, and no wings. Moreover, there are internal differences as well.

Classes are divided into orders. Among insects, the order Hymenopter includes stinging insects and ants. The order Orthoptera includes grasshoppers and crickets, and the order Diptera includes flies.

Nothing we have discussed thus far is exhaustive. There are Arachnids other than spiders, and there are orders of insects other than the three just described. However, most people consider everything among the Arthropoda as simply bugs, so we will switch to a more charismatic phylum for the rest of the discussion. By the way, true bugs, to be specific, are all within an order of insects called Hemiptera.

The phylum Chordata includes all animals with bones, ourselves and most of the animals with which we are most familiar among them. We and other Chordates, who have hair and bear living young, are in the class Mammalia. If we consider the order Carnivora among the mammals, we are talking about animals with long canine teeth, sometimes referred to as fangs. Animals in this order include some of our favorite pets, such as dogs, cats, and ferrets. However, it also includes the large, predacious animals that for much of human history were serious threats to our ancestors' well being. Within orders are families, and one family in particular within the Carnivora is the Felidae, the cats. If we talk about the family Canidae within the Carnivora, we would, of course, be referring to dogs. But at these levels, differences still exist that keep one type of animal in either, or any for that matter, family genetically isolated from others. Very few people would argue, for example, that lions and housecats are related, but practically nobody would suggest that one could possibly mate with the other. Indeed, the very thought is absurd. Besides the obvious difference in size, the two animals fall into different divisions within the family known as Genera (single: Genus). Lions are in the Genus *Panthera,* while housecats are in the genus *Felis*. In general, members of the genus *Felis* are small cats. Some biologists include the mountain lion within that genus, because mountain lions can purr. Other large cats cannot. More recently, however, some biologists have split the mountain lion into a different Genus: *Puma*.

Species are divisions within Genera. It is within the species level that evolution is said to occur, and it is the splitting of one species into two, what we speak of as speciation, that accounts for the basic divergence that ultimately accounts for biologic diversity. Moreover, we can finally define species, which would be a collection of organisms that resemble one another, will mate with one another if given the opportunity, and will produce fertile offspring. Another way of defining a species is to call it a collection of organisms that share a common gene pool. Thus, lions, members of the species *Panthera leo* are all capable of mating with one another, but normally not with tigers, *Panthera tigris*.

There is one other characteristic of species. All members of a given species typically share an identical number of chromosomes. Chromosomes are long strands of DNA that are found within the nucleus of cells. They contain the genes that are responsible for hereditary traits, which will be discussed in detail in the following chapter. Chromosome number is a species-specific characteristic, and it is perhaps the most functional feature that keeps separate species from blending. For example, horses typically have 64 chromosomes in each of their cells, and donkeys have 62. Horses and donkeys do resemble one another, and horses and donkeys can be induced to interbreed. When members of separate species successfully interbreed, the offspring is called a hybrid. The hybrid of a male donkey and a female horse or mare is called a mule, an animal with 63 chromosomes per cell that has been invaluable in the American West prior to the development and widespread use of motorized vehicles. Some people consider a mule to be a superior animal to either of its parents, but it has one distinct disadvantage to both: mules are typically sterile. With the exception of a few females, mules are incapable of breeding.

Lions and tigers also resemble each other somewhat. Both are clearly cats, but few people would argue that they are separate species. At one time, the lion was much more widely distributed than it is today, with populations that ranged through Europe and southwestern Asia, and even through the western Americas. A remnant population in the Gir Forest of India is all that is left of lions outside of Africa. Undoubtedly lions' and tigers' ranges overlapped, but it is extremely unlikely that any natural interbreeding ever occurred, because the animals have different habitat preferences and mating behaviors. However, lions and tigers share an equal number of chromosomes in their cells, 38, and crosses between the two species have occurred in captivity, comments made earlier notwithstanding. The offspring of these crosses, known as a tiglon if the male parent was a tiger or liger if the male parent was a lion, may be fertile, particularly the females if they are back-crossed to a male of one of the parental species. It is possible that they will produce fertile young. Male fertility of the hybrid is apparently another issue, however, which is consistent with the argument that lions and tigers are indeed members of different species. In the case of other animals, the subject is not always so clear.

Both the domestic dog and the timber wolf fall within the species *Canis lupus*. This, in fact, is the genus and species name of wolves, per se. Though once considered a separate species, dogs are now considered to be a subspecies of wolves and are thus designated as *Canis lupus familiaris*. This makes sense, because dogs and wolves easily interbreed, and deliberate crosses have been made, in part in attempts to breed aggressive but still manageable animals. Crosses are most easily accomplished between wolves and larger, wolf-like dogs. Crosses between wolves and something like a Shih Tzu, while theoretically possible, are unlikely, if not ridiculous.

Domestic dogs are an enormously diverse group of animals. Even though a cross between a Great Dane and a Yorkshire terrier are again theoretically possible, the idea of the two mating seems at least as absurd as the wolf and the Shih Tzu. However, accepting the definition of a species as a collection of organisms that share a common gene pool, the idea becomes more realistic, because a mutation that occurs in a Yorkshire terrier could theoretically make its way through the entire dog population, including great Danes. Admittedly, it would take many intermediate crosses and a large number of generations, but the possibility does exist.

Hybridization has also been known to occur between dogs and North American coyotes (*Canis latrans*), and fertile pups usually result. In fact, hybrids were known to have occurred naturally when coyotes were expanding their range into the upper Midwest and Northeast of the United States during the last third of the twentieth Century. Now that coyotes have established themselves in these states, coydogs, as the hybrid is known, are less common, suggesting that coyotes prefer to mate with members of their own species rather than dogs. Moreover, while wolfdogs appear to be genetically vigorous, coydogs are not. Behavioral and physiologic differences between coyotes and dogs result in offspring that really do not fit into either species particularly neatly and after a few generations produce somewhat sickly animals. Consequently, the separation of dogs and coyotes into distinct species would seem warranted, although the genetic divergence probably did not occur all that long ago in biologic history.

Hybridization in nature is probably a rare event. When it occurs, there has probably been some human involvement. For example, aquarium hobbyists who have mixed live bearing swordtail fish (*Xiphophorus helleri*) and platyfish (*Xiphophorus maculates*) have sometimes found that the two will hybridize. Likewise rainbow trout (*Onchorynchus mykiss*) will readily hybridize with cutthroat trout (*Onchorynchus clarkii*) when they find themselves in the same waters. Consequently, the stocking of rainbow trout into rivers and lakes that originally held only cutthroats has resulted in rainbow/cutthroat hybrids. From an evolutionary view, then, one would be forced to ask whether the two are indeed separate species or whether they are subspecies of the same species, perhaps the descendants of populations that became reproductively separated at some point in the past but too recently for complete speciation to have occurred.

That sort of question presents an interesting intellectual exercise for evolutionists, but it also presents something of a problem for fisheries managers. A species of trout unique to the southwestern United States occurs in two distinct subspecies: the Apache trout (*Onchorynchus gilae apache*) and the Gila trout (*Onchorynchus gilae gilae*). The southwestern US is largely desert, and trout habitat is at a premium. Consequently, the ranges of the two varieties of fish are quite restricted. Both cutthroat and rainbow trout have been introduced into waters inhabited by both varieties, and hybridization has occurred. Again, this raises the question about whether or not the fish actually belong in separate species, but it also endangers the Apache and Gila trout. They could literally be bred into extinction. Trout fishermen are often willing to travel in order to catch what they consider to be "exotic" species, and they spend money at their destinations. If the exotic species were no longer available, they would be less willing to travel to and spend money at that location, particularly if it is to fish for species they have available closer to home. But once again, the question about whether or not these trout are indeed separate species from the cutthroats and rainbows applies.

By now, most people reading this will have noticed that genus and species names are italicized, and the genus name is capitalized while the species name is not. This is an international convention probably based on another convention that genus and species names are usually from the Latin. Even when they are not, they are "Latinized." The advantage to this is that Latin, because it is a dead language, does not change over time. Spoken languages tend to have an evolution of their own. For example, American English and Australian English both originated from a common King's English dialect. But over time, each of the descendant languages has developed idioms of their own, and it is conceivable that given enough time with enough isolation, each dialect could evolve into a separate language.

Another convention is that a species name generally is given essentially as a modifier, while a genus name identifies an organism. For example, the species name *americanus* is used to designate an organism found somewhere in the Americas, but it tells nothing else about the organism. Seeing the name used by itself, one would not know whether it was designating a black bear (*Ursus americanus*), a redroot plant (*Ceanothus americanus*), a giant waterbug (*Lethocerus americanus*), or something else. In contrast, the genus name *Ursus* specifically identifies a bear. The species name is never given without a genus name preceding it. In the case of a very familiar organism, or one that has recently been mentioned in a printed document, the genus name is often represented by its initial. For example, the common bacterium often identified as simply *E. coli* is in reality *Escherichia coli*.

As one descends the hierarchy from kingdom to species, one is traveling from the broad and general to the narrow and specific. It is also from the most inclusive to the most restrictive. All members of any given species are members of the same genus. All species within a given genus are members of the same family, and so forth. Consequently, by studying this hierarchy, one is also studying relationships between organisms. The more categories a pair of organisms share, the more closely they are related, and the greater the evolutionary history that they have in common.

The organization of organisms into the categories described above, or more properly the classification of organisms, is the branch of biology known as taxonomy. Humans have been classifying organisms for our entire existence. Our original scheme was probably practical: what animals were dangerous; which animals and plants were edible, and the like. The systematic scheme that we employ now was originated by the Swedish physician Carl von Linné, also known as Carolus Linnaeus, in 1735. von Linné most likely organized his taxonomy on the appearances of the organisms he studied, assuming that the more superficially similar two organisms happened to be, the more closely they were related. We continue to use his system, although we have made many changes to his original derivations.

von Linné did not have the advantages we enjoy today. He did not know about DNA, let alone have the equipment to study it. Consequently, there is no way he could have known about how it can be used to classify organisms. Indeed, the microscopes that were available to him were too crude to enable one to see, let alone count, chromosomes. Thus, superficial similarity was a logical basis for classification. However, it did lead to some errors. Perhaps the most glaring of this was the misclassification of us.

When von Linné was putting his classification scheme together, he may not have known about the existence of great apes. Indeed, few Europeans knew about them, and those who did had major misconceptions. Among the misconceptions that existed well into the twentieth century was that all of the great apes were more closely related to one another than any one of them was to us. It was a reasonable error. After all, they all resemble one another more than any of them resembles us. Moreover, each of the great apes have 48 chromosomes per cell, or 24 chromosome pairs, which is discussed in the following chapter, while humans have only 46. However, once techniques for analyzing DNA were developed, it was discovered that chimpanzee and bonobo DNA are more like human DNA than like that of any of the other apes, and human chromosomes line up with chimpanzee chromosomes almost perfectly, despite the difference in chromosome number. It turns out that human chromosome number two closely matches two chimpanzee chromosomes, designated 2p and 2q, if they are laid end to end. It is thought that sometime in our biological history after our and the chimp ancestors behaviorally parted ways, two chromosomes fused end to end (IJdo et al 1991). In support of this idea is an oddity in human chromosome number two. All chromosomes have units of repeating DNA called telomeres at their ends. However, human chromosome number two has vestiges of telomeres more or less in its middle as well, and the chromosome grossly resembles two sausage links hooked together. Moreover, all chromosomes have a structure called a centromere by which they attach to fibers in a cell during cellular reproduction. Human chromosome number two has a remnant of a second centromere in one of its links. This phenomenon has undoubtedly occurred in other organisms. Horses and donkeys, for example, appear to be closely related, which suggests a common ancestry, and are separated from one another by a single chromosome pair. The same argument could be made for elephants and wooly mammoths. Sometimes one finds chromosome number variation within a species, as with both cutthroat and rainbow trout. But DNA similarity is the critical factor, and all members of a given species share similar DNA. Moreover, the more closely two organisms in separate species are related, the more similar their DNA will be.

As mentioned early in this chapter, it is at the species level that evolution occurs. Whenever a population is fragmented and barriers to reproduction between subpopulations develop for a long enough period of time, the splintered part of the population will ultimately diverge from the parental part to the point that if the barriers are removed, interbreeding will have become impossible. In essence, a new species has evolved, or one could say that speciation has occurred. The end result is an increase in biodiversity.

Reference

IJdo JW, Baldini A, Ward DC, Reeders ST, Wells RA (1991) Origin of human chromosome 2: an ancestral telomere–telomere fusion. Proc Natl Acad Sci U S A 88:9051–9055

Chapter 10
How does it Work?

Imagine that you own an island that is inhabited by a population of gray rabbits. Imagine further that one day you discover a tan rabbit on the island, and for some reason you prefer it to the gray ones. You start trapping rabbits and get rid of all of the gray ones you catch. The occasional tan one that shows up in one of your traps is released back onto the island. At first tan rabbits will show up in your traps only rarely, and they will be observed infrequently. Over time, however, as you selectively remove the gray ones, the tan ones will become more and more common, if there is nothing selectively eliminating them. If you continue eliminating the gray rabbits, the tan ones will eventually become more numerous, and if you live long enough, you may succeed in reducing the gray rabbits to an occasional individual only. You may not eradicate them, but you will have succeeded in changing the population of rabbits from predominately gray to predominately tan.

As has been mentioned several times up to this point, natural selection works in the same manner. All other things being equal, an organism that carries a trait that is favorable in a particular environment is more likely to survive and reproduce successfully than an organism of the same species that does not carry that trait or, worse, carries a trait that puts it at some kind of specific disadvantage.

The situation described in the opening paragraph is obviously artificial, but similar events occur often enough in the real world, where selection occurs either by breeders or by nature. Sometimes selection occurs inadvertently, as in the case of the superbugs described in Chap. 6. Often it is deliberate, and sometimes it is accidental. Aquarium goldfish were probably first domesticated in China around a millennium ago when people discovered brightly colored mutants among otherwise drab fish in the wild. The brightly colored mutants most likely appeared periodically in wild populations, but they never became common because their bright color stood out against the dark background normally found in the kinds of bodies of water that goldfish prefer, and predators had an easier time seeing them. Thus, they were naturally eliminated from wild populations. Once brought under domestication, they were protected and selectively bred, and today most people outside of biology are

B. Marcus, *Evolution That Anyone Can Understand*,
SpringerBriefs in Evolutionary Biology, DOI: 10.1007/978-1-4419-6126-6_10,

unaware that the natural color of goldfish is drab brown. Contemporary goldfish breeders often find drab brown individuals showing up among the offspring of their brightly colored specimens. Goldfish are able to tolerate water with low levels of oxygen, often surviving where most other fish cannot, and people who have wanted to get rid of pet goldfish humanely have often released them into creeks and ponds. In water in which the goldfish can survive but a predator like a bass cannot, brightly colored goldfish may reproduce, although they will probably have drab brown individuals among their offspring. Thus, in bodies of water where self-sustaining goldfish populations now exist, thanks to the misguided efforts of well intended people, one can find brown specimens among the orange, yellow, white, and multicolored ones. But if predators are present, the brightly colored goldfish becomes appealing prey. Consequently, between domestic and feral goldfish populations, one can see both natural and artificial selection at work.

All of this may sound theoretically possible, but to many people who have no direct experience with this sort of thing, some concrete proof would make it more acceptable. Such proof is abundant. In fact, in the years between framing and publishing his theory of evolution, Darwin himself collected a considerable amount of it.

Darwin was well acquainted with domestic animals and plants, and he knew that many varieties or breeds of them existed. In addition, he became quite interested in the breeding of pigeons, of which several varieties exist. In learning about them he found reinforcement for his theory. He also studied dogs. Darwin wrote about this in *The Variation of Plants and Animals Under Domestication* (Darwin 1868).

More recent proof, and perhaps more compelling, was provided by the Russian biologist Dmitry Konstantinovich Belyaev. In 1959 in Novosibirsk in Siberia in what was then the Soviet Union, Belyaev began a series of experiments in which he attempted to breed tame silver foxes, a color variant of the red fox *Vulpes vulpes*. He did this by starting with wild animals but selecting for the least fierce among them. He repeated this over several generations, always selecting those that responded to him least negatively for further breeding. In time, he succeeded in producing foxes that were more and more dog-like in behavior. Not only did the animals become friendly, but they also developed dog-like characteristics of barking and wagging their tails, developed color patterns like dogs, and they lost their always erect ears. (Trut 1999) The experiment continues in Russia today, and at the very least it shows that animal forms are not immutable.

Again, one can argue that such evidence is tainted because it occurred under artificial conditions. Can it be shown to have ever occurred in nature? It can.

The peppered moth, *Biston betularia*, is a nocturnal species that spends its days resting on trees. In the past in England it was predominately mottled gray in color, and the trees it generally preferred to rest on were encrusted with lichens that were similar to the moth in color. Moreover, the lichens reflected ultraviolet light more or less similar to the moth. Dark or melanistic specimens of peppered moths occur, but these stand out against the lichens when the insects are resting, and individuals with dark coloration are quickly picked up and eaten by birds. Lichens are highly sensitive to air quality, and when the industrial revolution began polluting the air in England, the lichens began to die. That left the tree limbs dark. Moths continued to rest on trees by day, but the gray ones were now more visible against the darker bark, and birds began selecting them rather than the melanistic moths. As a result, the population of moths began shifting from gray to dark. When air quality control laws were passed during the latter half of the twentieth Century, and air quality improved, lichens again began growing on the trees, and the lighter moths were once again at a selective advantage. Thus, where a mutation confers no physiological or behavioral advantage on an organism, it may offer an environmental one.

A question that may have occurred to a number of readers by this time is: where did that first tan rabbit or first brightly colored goldfish come from? Similarly, what accounts for the differences in temperament, however slight, among wild foxes or in color among moths? The answer to these questions is, of course, changes in genes, what we refer to as mutations.

The first systematic study of the inheritance of gene mutation was conducted by an Austrian monk named Gregor Mendel. Mendel was well educated and a keen observer. It is said that in the garden of the monastery at which he lived, he noticed that pea plants in different parts of the garden showed a variety of traits that always seemed to be passed consistently from one generation to the next. For example, plants with green pods always produced offsprings with green pods, and plants with yellow pods always produced offsprings with yellow pods. This piqued Mendel's curiosity. He tried passing the pollen from a plant with green pods to one with yellow pods, and he collected the seeds that were produced. When the seeds germinated, he found that the plants had produced green pods.

Had Mendel been satisfied with that result, he might have become at best little more than a footnote, if that, in the history of biology. However, he let that second generation breed and he again collected the seeds. This time, he found their offspring to be a mixture of plants with green pods and plants with yellow pods, with green outnumbering yellow by a ratio of around three to one. From these results, Mendel reasoned that within the pollen and the pollen-receptive cell of the flower there are "factors" that govern the transmission of traits from one generation to the next. Furthermore, he reasoned that when factors for two conflicting traits unite in a single organism, one of them, the dominant factor, would be expressed. Further still, he reasoned that the unexpressed, or recessive, factor does not disappear; it simply lies hidden and under the right conditions can be expressed in a later generation. From this we have what has been called Mendel's Law of Dominance.

We now know that Mendel's factors are the genes, the units of DNA that control heredity. We refer to genes that dictate different expressions of a given trait, such as green or yellow pea pods, as alleles. Thus, in the simplest state, there are alleles for dominant and recessive traits. Usually there are multiple alleles with varying degrees of dominance for any trait.

Genes are carried on cellular structures called chromosomes, which are found within the nucleus of the cell. In reality, the chromosomes are long strands of DNA, and the individual genes are sections of the strands that code for specific proteins. Not all DNA codes for proteins. In sexually reproducing organisms, that is organisms in which two partners must be involved in reproduction, chromosomes normally occur in pairs, with one member of each pair having been provided by each parent. Diagrammatically, a cell nucleus would look something like the following diagram:

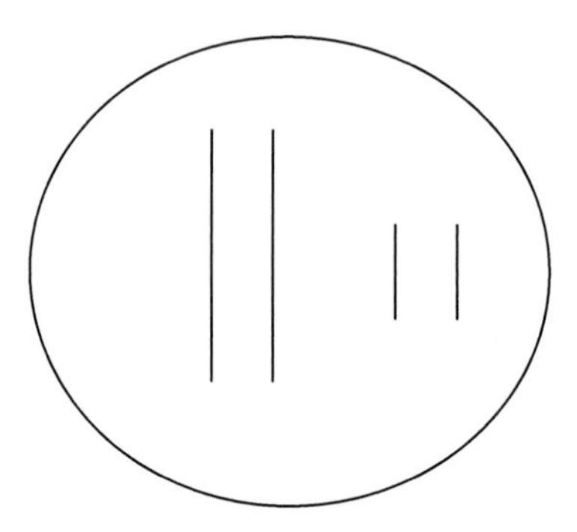

The preceding diagram would illustrate a cell with two pairs of chromosomes. Most organisms have many more. Humans, for example, have 23 pairs of chromosomes, and horses have 32.

When such organisms produce their own reproductive cells, the members of each chromosome pair separate from one another. Thus, each reproductive cell ends up with half the chromosome number of a regular cell. Again diagrammatically, it would look something like this.

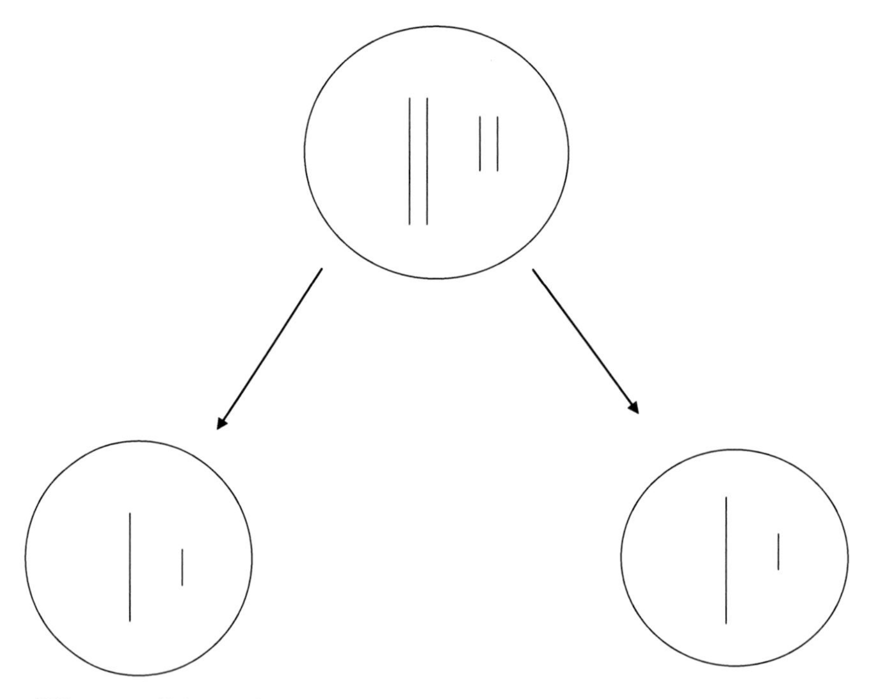

Whatever alleles each member of each chromosome pair is carrying go with it. This has been called Mendel's Law of Segregation. In this manner, chromosome number is maintained from one generation to the next. This is not the case when a cell simply reproduces itself, rather than producing a reproductive cell. In that situation, each chromosome makes a copy of itself, thus insuring that every cell, other than reproductive cells, has a copy of every gene an organism inherited. In both cases of cellular reproduction, unless an error occurs in copying, a mutation, each cell is virtually a copy of the one that produced it. On the whole, errors in copying are rare, but when one considers the number of copies that must be made, perhaps billions in a mouse for example, errors become virtually inevitable.

Mendel carried on his studies with peas for approximately seven years. He discovered that the inheritance of one trait need not influence the inheritance of another. For example, if the plant with green pods also had white flowers, and the plant with yellow pods had purple flowers, the cross between the two would have produced plants with purple flowers, as purple is dominant to white. Crosses between two hybrid plants would produce some offspring with purple flowers and some with white, again in a ratio approximating three to one. However, the color of the flowers has nothing to do with the color of the pods, and some third generation plants with green pods would have white flowers, as was the case with the original parental plant, and some would have purple. The traits are inherited independently, which has become known as Mendel's Law of Independent Assortment.

At this point, I would like to take a digression and discuss the inheritance of the trait that is necessary for reproduction to occur: sex. In mammals and most other sexually reproducing animals, sex is determined by a specific chromosome pair. In females, typically, both members of this pair grossly resemble the letter X, while in males there is one X chromosome and one that grossly resembles the letter Y. Thus, females are often designated as XX, while males are designated XY. Consequently, when reproductive cells are produced, the two sex-determining chromosomes segregate from one another and end up in separate cells, as do all other chromosome pair members. In the case of females, the process would look like this:

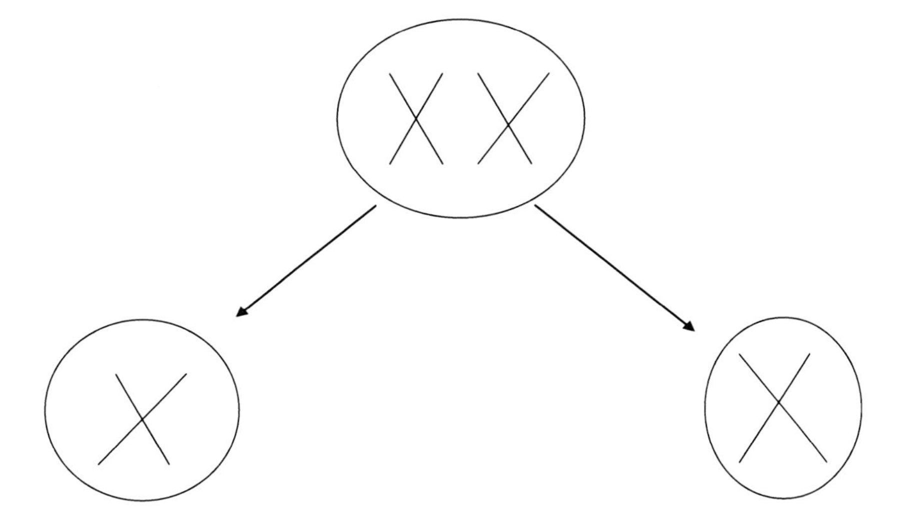

All female reproductive cells contain a single X chromosome. In males it would be

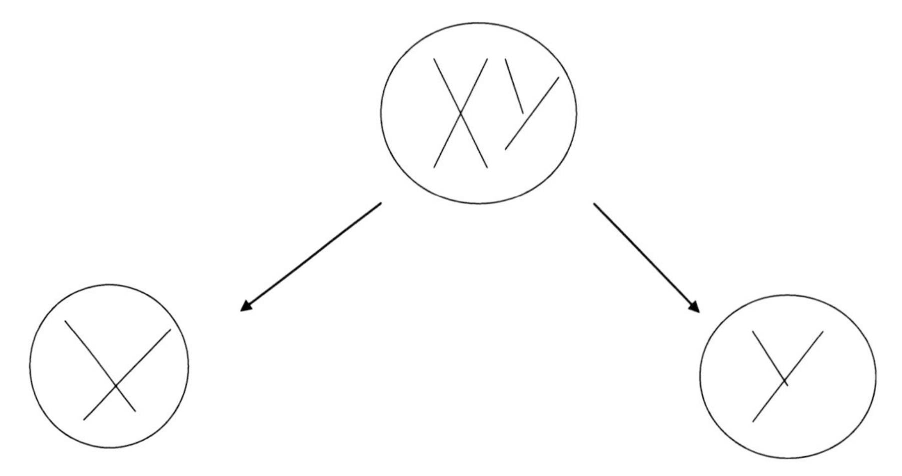

If a male reproductive cell, a sperm, carrying an X chromosome unites with or fertilizes a female reproductive cell, an ovum, the resulting individual will be female; if a Y-bearing sperm fertilizes an ovum, the result will be male. It is the male parent that determines the sex of the offspring.

Not all animals have XX/XY sex determination. In birds, for example, it is the female that determines the sex of the offspring, as she has two dissimilar chromosomes for sex, while the male has two identical ones. Among the hymenoptera, the order of insects that includes the ants, bees, and wasps, all females are XX, and all males are simply X. Males develop from unfertilized ova, a phenomenon called parthenogenesis. Among some other insects, such as the order that includes grasshoppers, males develop from fertilized eggs, as do females, but males have only a single sex-determining chromosome, while females have two.

In reptiles, which are ancestral to both mammals and birds, there are several mechanisms by which sex is determined. Some reptiles are like birds, where the male carries identical chromosomes. Others are like mammals, where the female carries identical chromosomes. In others still, sex is environmentally determined, specifically by temperature. Among some turtles, embryos in eggs that develop at warmer temperatures become females, while those developing at cooler temperatures become male. Among crocodilians, both higher and lower nest temperatures produce females, while intermediate ones produce males. Even more complicated are individuals in some frog and fish species that appear to change sex. In some reef fish, specifically, schools occur where all of the members are female except for one. If something happens to that single male, if it dies or is eaten by a predator, the largest female changes into a male. Among other fish, the exact opposite occurs. In others still, individuals may change sex if the population is severely skewed toward one sex or another. A female aquarium guppy, for example, may become male if there are only females in the tank.

The plasticity of some species with regard to sex suggests that chromosomes may not be the only determining factor. Instead, sex may sometimes be a matter of genes that respond to specific environmental signals. Moreover, that there are multiple means by which sex is determined shows that sex determination originated more than once and that there is more than one way to accomplish a biological end. It also underscores the importance of sex in reproduction. Some organisms are capable of reproducing without sex. A kind of waterflea, for example, is almost always female. Eggs are produced, but they develop and hatch without benefit of fertilization. Mutation is the only source of variability within the population, and the offspring produced are essentially clones of their mother. In a static environment to which the animal is well adapted, this may be of benefit. However, if the environment were to change, the entire population could be wiped out. In contrast, among animals, and plants, where reproduction is a result of sex, mutation is again the only source of variability, but unions of sperm cells and ova provide new combinations of genes, which often interact with one another. Such combinations enhance the variability of the organisms in question, and variability of a species is key to its survival in a changing environment. Over many generations, such variability can end up producing a vastly different species.

Just as sex determination is not carved in stone, neither are Mendel's Laws. In the case of dominance and human blood type, for example, both types A and B are dominant to O, but neither is dominant to the other. This means that someone with, say, type A blood, who is married to someone with type B, may very well end up with children who are AB. Likewise, someone with straight hair married to someone with curly hair may have children with wavy hair. Conflicting traits may both be expressed, or they may be blended, depending upon the actions of the genes involved. Many human traits are affected by combination of alleles rather than by a single allele pair, and identifying a specific dominant or recessive trait can be all but impossible. In other cases, there are genes governing other traits that may have to be involved. In the case of human Type II diabetes, for example, being overweight is a clear risk factor. However, not everybody who is overweight develops Type II diabetes. The necessary genes must be present as well. Unfortunately, they often are, but the example along with the others cited in this section reinforces the fact that not all hereditary traits are governed simply by a, single gene pair.

Other genetic factors influence the survival of a trait and, consequently, the survival of an individual organism carrying it. Some, known as lethal genes, may even bring about the death of an organism carrying it. An example of that is a trait called sickle-cell disease, which affects the shape of the hemoglobin molecule in humans.

Hemoglobin is the red pigment found within blood cells that carries oxygen from the lungs to the tissues. The shape of the molecule is determined by genes, and a specific mutation can cause the production of somewhat distorted hemoglobin. If a person inherits one such mutated allele, he is normally able to tolerate it. His life may be slightly shortened, and he may be unable to live at high altitudes where the air is thin, but the condition will not kill him. If he inherits two sickle-cell alleles, however, one from each parent, his quality of life is diminished markedly. Included among the symptoms are pain, fatigue, skin ulcerations, and jaundice. In most of human history, people with the double dose of sickle-cell disease often died in young adulthood. Today, at least, there are medical treatments that ease the symptoms and prolong life. However, people who have received only one allele for sickle-cell disease do have one advantage over those who do not have the disease; they do not get malaria. The parasite that causes malaria lives inside the red blood cell and feeds on hemoglobin. However, it is not able to use sickle-cell hemoglobin. In other words, someone who inherits a sickle-cell allele from only one parent, while being spared the worst of the disease, is at a selective advantage over someone who does not carry the allele. Consequently, the sickle-cell allele is quite common in areas where malaria is a problem, particularly in Africa, and it is very likely to persist.

To reiterate a point made earlier, none of Mendel's laws is carved in stone.

Also as mentioned earlier, genes are sections of DNA found in the nuclei of cells that code for proteins. This means that a series of biochemical steps starting with the genes ultimately ends in the production of a protein, such as hemoglobin, and the protein usually accounts for the expression of the gene. Proteins are long chains of compounds known as amino acids. There are 20 of these compounds that are involved in the construction of proteins, with the exact nature of the protein being a factor of how many of each amino acid is present and in what order they occur. DNA itself is a long chain of compounds called nucleotides, of which there are normally four, and every three of them codes for a specific amino acid. Thus, the sequence of amino acids in a protein is a product of the sequence of nucleotides in DNA. Any alteration in the nucleotide sequence, which is basically the description of a mutation, brings about an alteration in a protein, the result of which can be a different appearance or function of a trait governed by that protein. The molecular mechanisms that are involved in these processes are fascinating, but they are outside of the range and scope of this book. There are, however, quite a number of books on the subject that are written for nonscientists, and anyone interested in pursuing the subject in any of them is strongly encouraged to.

In general, changed traits resulting from mutations can affect the organism carrying them in one of three ways: it can benefit them, it can leave them unaffected, or it can harm them. The last possibility is probably the most frequent, often because a gene mutation may interrupt a biochemical pathway necessary for a specific function. Often a deleterious mutation is recessive to the normal or so-called wild-type allele, and an organism must inherit the mutated allele from both parents in order to express the trait. Again, sickle-cell disease would serve as an example. However, not all cases offer an advantage to the hybrid, as does sickle-cell disease. For example, the Manx cat that we introduced in Chap. 9 is originally from the Isle of Man, an island in the Irish Sea. The Manx has a mutation that affects development of the spine such that Manx cats have vestigial tails. The tails are often simply stumps, giving the cat the appearance of having its rear half look something like that of a rabbit. Indeed, some Manx cats even hop. When Manx cats mate, some of their kittens are born fully tailed, meaning that Manx cats are hybrids. They carry alleles for normal tails. No Manx cats with alleles for the vestigial tail only exist, probably because the allele causes spinal deformation that is lethal.

There are many alleles similar in nature to Manx. In humans, a mutation that affects a chemical pathway in nervous tissue does not appear to do major harm. However, in people with two alleles for this condition, the usual outcome is death by age of five or earlier. Known as Tay-Sachs disease, this malady is lethal to the person who inherits it from both parents.

Mutations that do not appear to affect an organism are common as well. A black coat color occurs in gray squirrels in parts of Upstate New York and southern Ontario. The trait appears to be nothing more than a variant in the more typical coloration of the animals and appears to be immaterial in terms of a squirrel's fitness.

As for beneficial mutations, that would depend somewhat on your point of view. Few of us would consider the mutations that rendered pathogenic bacteria unaffected by antibiotics or mosquitoes resistant to DDT as beneficial, but the bacteria and mosquitoes certainly would have were they capable of it. In humans, we know of mutations that render some people resistant to HIV infection and mutations that allow some people to carry high levels of blood cholesterol with no ill effect. Such would be considered to be beneficial.

Many mutations cannot be arbitrarily described as harmful, beneficial, or neutral. It is the environment that makes the determination. One of the classical examples of this is the industrial melanism in peppered moths mentioned earlier in this chapter. It is a case where a mutation confers no physiological or behavioral advantage on an organism, the environment determines whether or not a morphological one will.

When Darwin and his predecessors began thinking about evolution, they had no way of knowing that the external traits they examined were determined by internal biochemical mechanisms. Even Mendel, who postulated the existence of genes, had no idea how they worked. In contrast, when Dmitry Konstantinovich Belyaev began his experiments on fox domestication, he knew very well that external traits were governed by internal, biochemical mechanisms. Indeed, part of his goal was to demonstrate that domestication brought about changes in those mechanisms.

The preceding discussion barely scratches the surface on the science of genetics. Even Mendel's Laws are described in the scantest of detail. But it is important that one appreciates that genetic change is very much a reality of living things, and such change can affect how an organism interrelates with its environment. Moreover, the idea of the environment selecting what survives and what does not is the essence of natural selection, and it has been described throughout this book. It is fundamental to anything involving evolution, including changes in molecular biology.

References

Darwin C (1868) The variation of plants and animals under domestication. Available from The Johns Hopkins University Press (1998), Baltimore

Trut LN (1999) Early canid domestication: the farm-fox experiment. Am Sci 87:160–169

Chapter 11
What is the Evidence?

Less than 10 min by car from Beverly Hills, when the traffic is cooperative, on Wilshire Boulevard in Los Angeles, you will find the Los Angeles County Museum of Art and George C. Page Park. In the park you will find the Page Museum and a large, somewhat foul smelling pond with statues of mastodons around and in it. This is the La Brea Tar Pits, a bed of asphalt right under the ground surface formed from oil seeping up from below over many thousands of years. In some places water has collected over depressions in the asphalt. Animals that have approached the water to drink have become entrapped in the asphalt. Others chased into or simply wandering through the area have sunk through the shallow layer of ground above the asphalt and likewise become entrapped. Predators that attempted to capture the trapped animals also became mired in the ooze. As the animals died and sank, their bones became preserved, and today the La Brea Tar Pits have become one of the most important sources of fossils in the United States. This in downtown Los Angeles.

Among the animals fossilized in the tar pits are several mentioned earlier. They include giant ground sloths, glyptodonts, and saber-toothed cats. In addition there are American lions, short-faced bears, dire wolves, and mastodons, all animals that nobody has seen wandering around Southern California recently. These, in fact, are animals of the last ice age, ones that disappeared after the arrival of humans in the area. Oddly, only one set of human remains have been retrieved from the pits thus far. That is the partial skeleton of a young woman, whose skull had been crushed, suggesting that she had been killed and then flung into the tar pits.

The fossilized bones that have been retrieved from the tar are recent, in terms of geological time, and are well preserved. Similarly, remains of animals like wooly mammoths have been retrieved from glaciers. Presumably the mammoths sank in deep snow, died, and froze as the snow was compressed into ice. Such remains are in good enough condition that ideas of extracting cell nuclei from their tissues and, using cloning technology, trying to implant an embryo into the uterus of an elephant have been floating about.

B. Marcus, *Evolution That Anyone Can Understand*,
SpringerBriefs in Evolutionary Biology, DOI: 10.1007/978-1-4419-6126-6_11,

Fossils include virtually anything that is evidence of past life. For the most part, people think of fossils as animal hard parts like bones, teeth, or shells that have somehow become preserved in rock. Indeed, most of our oldest fossils are stone, and they have been studied extensively. In fact, it was his knowledge of geology and his having studied fossils that contributed to Darwin's framing his theory of evolution.

As indicated above, it is usually an animal's hard parts that get preserved, and the process is not easy. Older fossils are typically found in sedimentary rock, that is rock that has formed from fine debris like sand or silt settling from air or water and then become compressed as more and more debris settled on top of it. Any plant or animal or their remains can become buried. Before the accumulating material is compressed into stone, water percolating through it may dissolve away the organic remains of any life form that is within in, only to replace it with minerals it is carrying. The result is that the organic material is replaced by inorganic, and a cast forms that is essentially a model of the organism that was there. Over time petrifaction occurs and a fossil is formed. Another possibility is that a carbon imprint of the life form persists as the sedimentary material is compressed into rock. In either case, there is a remnant of past life.

The accumulation of sediment would at first seem to be very, very, slow, and it is difficult to imagine that something that has died would not decay long before it was buried. In reality, there are cases where sediment accumulation can occur quite rapidly. Rivers, for example, often carry substantial loads of sediment in the form of sand, silt, and clay. Where a river flows into a large body of standing water, like a large lake or the ocean, the current slows and the sediment settles, with sand going first and clay last. Even more rapid depositions occur as a result of catastrophes, such as a volcano erupting, a landslide, or an earthquake. An erupting volcano can deposit tons of ash in a very short period of time, and entire mountaintops can be dislodged by a landslide or earthquake, only to bury whatever might be at the base of their slopes. One particular example of this is thought to have occurred more than half a billion years ago in what is now Mt. Field in Yoho National Park in British Columbia, Canada. In a ridge of rock known as The Burgess Shale, one can find a rich deposit of a variety of fossils, some of which are clearly ancestral to organisms alive today, and some of which are unique to that particular time in geological history (Gould 1990).

Many fossils can be used as time indicators. For example, a rock bed that is full of fossil *Trilobites* would indicate that the rock formed during the Paleozoic Era. That, after all, is when *Trilobites* existed. One would not find fossil mammal bones among *Trilobite* fossils. By the same token, human and dinosaur bones are not found together. Dinosaurs went extinct long before humans originated. That being said, parenthetically, a search on the Internet will bring up a number of sites claiming that humans and dinosaurs coexisted, and many of those claim to offer scientific proof. Some of those, and others, further claim that scientists are conspiratorially keeping news of coexisting human and dinosaur fossils from the public in general. In general these sites are religiously, rather than scientifically, oriented, and the proof they offer include references to the Bible, to creationist literature, and to antievolutionary bias and even conspiracy. One specific instance of human and dinosaur bones having been found together is the so-called Moab Man, supposedly a deposit of human and dinosaur bones found together in Utah. Examination of the human bones revealed that they were hundreds of years old, not millions, they were dehydrated but not petrified, and they most likely represent human remains having been deposited among dinosaur fossils, as perhaps during burial practices by Native Americans living in the area.

An explanation that has sometimes been offered for fossils is that they represent organisms that were killed in Noah's flood. Were that the case, how is one to account for fossil fish and amphibians? After all, these are aquatic animals well capable of swimming. Moreover, why are there such obvious strata of fossils? If fossils indeed represent the remains of animals that drowned all at once, there would be no order to them. Horse and dinosaur bones would be found together, which they are not. In addition, all ancient and modern forms would be found together, and they are not. Fossil-bearing rock formations can be dated using radiometric techniques, and the oldest of such rocks typically contain only invertebrates, animals without bones. They appeared before vertebrates did. Among the vertebrates, fossils of primitive fish appeared first, followed by amphibians and then reptiles.

Fossils of reptiles begin appearing in rocks between 310 and 320 million years old. Ancient reptiles, in turn, gave rise to modern reptiles, birds, mammals, and dinosaurs. Recently, the origin of dinosaurs has been pushed back to around 250 million years ago (Brusatte et al. 2011). Mammals originated around 50 million years later, but they remained small and inconspicuous for the next 140 million years, as the dinosaurs ruled the planet at the time. It was not until after the dinosaur extinction that mammals began to enlarge and diversify. Consequently, the only mammal bones found in sediments with dinosaur remains are small and mouse-like. If dinosaurs had died in Noah's flood along with so many larger mammals, their bones would be found together.

One of the criticisms that have often been hurled at using fossils as indicators of evolution is that the fossil record is incomplete; it is full of gaps. One often hears the term missing link used for fossils that are transitional between two sequential types. In fact, the fossil record is quite complete. Some gaps do exist, but over the years many of those so-called missing links have indeed been found, and the gaps have been filled in. In fact, some fossil sequences are quite complete.

Among the fossil lineages that show a good evolutionary record is that of whales. Recall that in Chap. 4 I mentioned that some whales have rudimentary leg bones, what was described as vestigial structures, which I cited as evidence that whales had terrestrial ancestors. A 52-million-year old fossil of one possible ancestor was discovered in Pakistan in 1978. The skull had ear structures that were intermediate between those of earliest whales, as gleaned from other fossils, and even older fossils of wolf-sized terrestrial carnivores. More recent fossils showed hind legs that were adapted for swimming. Fossils from around 40 million years ago showed even more aquatic adaptation, including hind limbs that could no longer function on land. While nobody can dogmatically claim that the fossils represent animals that were direct ancestors of modern whales, they do show that transitional forms do occur in the fossil record. Furthermore, the transition from land-dwelling to water-dwelling mammals has occurred more than once. Manatees are structurally similar to whales. Both have fore limbs that are modified into paddles for swimming, and both lack obvious hind limbs. However, the manatee's closest four-legged relative is the elephant, while the whale's is the hippopotamus.

The classic evolutionary line, with numerous intermediate fossils, is that of the horse. It was first worked out around 130 years ago, although there were some errors in the sequence. The paleontologist who did the work, O.C. Marsh of Yale University, used fossils to determine a pretty straight line between the modern horse and its earliest known ancestor, the beagle-sized *Hyracotherium* or, as it is often called, *Eohippus*. Since then, the discovery of more fossils has revealed that the pathway was in reality quite jagged with many blind ends. The modern horse does not happen to be the end of the sequence; it is simply a modern survivor, along with donkeys and zebras.

There are people who do not object to evolution in its entirety. Rather, they object to aspects of it. Many of these concede that horses, donkeys, and zebras evolved from a common ancestor, but they object to the idea that one class of animals could possibly give rise to another. They dispute that birds and mammals, for example, could be descendants of reptiles, and they often cite an absence of transitional fossils to support their position. And once again, the argument is fallacious.

In 1860, in 150-million-year old limestone deposits in Solnhofen in Germany, a single fossilized feather was found. In 1861, a small dinosaur fossil with sharp teeth and a long tail was discovered, a fossil of a dinosaur that was found to have wings and feathers. Animals resembling modern birds were not around 150 million years ago, but apparently feathered dinosaurs were. The fossil in question was named *Archaeopteryx*. It was not a bird. Rather it was more likely a theropod dinosaur, a member of a group with birdlike characteristics. Theropods walked on their hind legs; they did not fly. They had hollow bones; and they had wishbones. Some may have had feathers as well, which they probably used for insulation. *Archaeopteryx*, again, not a bird, appears to be at most descended from theropods and ancestral in form to birds.

The transition between fish and terrestrial vertebrates is another spectacular step in the history of life, and there are intermediate fossils that document the passage. Even more dramatically, there are living organisms that represent the transitional forms. My personal favorites are the lungfish, a group of primitive fish some of which are able to survive their ponds drying up by breathing air. Lungfish populations are found today in Australia, southern Africa, and South America. Fossils suggest that they were widely distributed at one time. The existing populations are surviving remnants.

Lungfish are true fish, but they are in a different classification than the fish most of us are familiar with. Common fish in North America, in both fresh and salt water, are what are called ray-finned fish, while lungfish are considered to be lobe-finned fish, an anatomic characteristic that makes them more closely related to four-limbed animals than are the ray-finned. Indeed, it is this anatomical similarity that suggests their relationship to land animals more than their capability of breathing air. Lungfish are very primitive; they resemble fish that were around more than 350 million years ago.

Yet another lobe-finned fish that survives today is the coelacanth. It was first found in the Indian Ocean off east Africa in 1938 after having been thought to be extinct since the end of the Mesozoic Era. Another population has since been discovered in the waters off Indonesia. Coelacanths are related to both lungfish and four-legged land animals, and while the existing two species of coelacanths are adapted to their particular lifestyles, they still strongly resemble the fossils that are in the line between fish and land animals. Moreover, the movements of their paired fins, the pectoral and pelvic, are curiously like the movement of the limbs of four-legged animals rather than being typical of fish. Parenthetically, coelacanth has sometimes been referred to as a Lazarus taxon, a category of living things that were thought to be extinct but then relict living populations have been discovered. The most recent example of such a thing might be the ivory-billed woodpecker, which used to live in the southeastern United States. The bird has been considered to be extinct since the middle of the twentieth century, but then a sighting and brief video recording were reported from Arkansas in 2004. Sadly, the existence of ivory-bills has not been confirmed, thus it is probably premature to classify the species as a Lazarus taxon. If anyone is wondering if the name Lazarus comes from the story in The Gospel of John 11:41–44, it probably does.

Finally there is the case of *Tiktaalik*, a lobed-finned fish that lived around 375 million years ago and probably resembled animals intermediate between fish and the first four-legged animals. There are no known living populations of this genus, and none are likely to be found. Fossil evidence suggests that *Tiktaalik* lived in shallow, probably oxygen-poor waters. The discovery of *Tiktaalik* fossils in 2004 was no accident or act of serendipity; the paleontologists who found them were out looking for them in a pretty inhospitable place: Ellesmere Island in Canada at the north end of Baffin Bay, well north of the Arctic Circle. They were looking in sediments of the right age and right composition, sediments in which the kind of fossil they were seeking should have been found.

One of the realities of scientific theories is that they can be used to make predictions. As such, the discoverers of *Tiktaalik* predicted the kind of sediments that would hold the type of fossil for which they were searching, found it, and then found the fossil. Things do not always work out that smoothly, a lot of hard work went into the actual find, but the fossil was where it should have been. By the way, *Tiktaalik* most likely did not walk on land. It was more probable that it propped itself up on its forelimbs or perhaps more appropriately its fore-fins, lifting its head above water level. But it is still representative of a life form intermediate between two others, and thus it must be added to the roster of missing links that in fact do exist in the fossil record.

In Chap. 4 we spoke about the Antarctic ice fish and its fossil genes. We also spoke about the vestigial leg bones in whales and constrictor snakes, a relict tail in human embryos, and other vestigial organs. These non-functional remnants tie in with what was once a principle of biology known as the recapitulation theory, expressed as "ontogeny recapitulates phylogeny," which essentially states as an animal embryo develops toward adulthood, it passes through stages that resemble its evolutionary history. Thus, a developing chicken, for example, would pass through stages where it resembled a fish, a reptile, and finally a bird. Contemporary biologists do not consider this as ironclad, but as a chicken embryo develops, it does go through stages with similarities to ancestral species. Early in its development, its dorsal aorta, the largest artery in the body, folds into a two-chambered heart and branches form off it to become gill arches. Its limbs resemble fins. These characters bear resemblance to those of fishes. Later on, the embryo shows a three-chambered heart, and the gill arches have become modified into arteries, such as the pulmonary arteries that transport blood to the lungs for carbon dioxide to be exchanged for oxygen. This is characteristic of reptiles. Finally, the heart develops a fourth chamber, which is characteristic of birds. Mammal auditory ossicles, ear bones, are generally believed to be descended from the lower jawbone of reptilian ancestors. Modern reptiles have several bones on either side of their lower jaws but only a single one in each of their middle ears. Mammals have a single bone on each side of the lower jaw but three in each middle ear. Mammal embryos go through a formative period where their ear bones are attached to their lower jaws but then separate as the animal develops. Parenthetically, a 125-million-year old fossil found in China shows a small mammal with its ear bones attached to its jaw (Bartlett 2007). This fossil, by the way, would qualify as a transitional fossil or missing link, but it serves to underscore the idea that as animals evolved, existing structures became modified to serve new functions. This did not happen toward some goal; it occurred randomly, but natural selection favored the modifications, and they persisted.

In most cases, plants and animals of related species tend to resemble one another. Thus, cutthroat trout and rainbow trout look a lot alike, as do red and sugar maple trees, silver and arctic foxes, and rock and mourning doves. A Swedish physician by name of Carl von Linné, known to us today as Carolus Linnaeus, mentioned earlier, published a classification of living things. It is probable that Linnaeus used superficial similarity as his basis of classification, thus he considered foxes and wolves closely related. Subsequent biologists followed suit. It is only reasonable that the more two species resemble one another, the more closely related they are likely to be. The discovery of DNA and eventual unraveling of genomes generally has supported this idea, and foxes and wolves are more biochemically similar to one another than either is to, say, a raccoon. However, similarity in appearance is no guarantee of close relationship, and dolphins, for example, despite their gross similarity to sharks, are not particularly closely related to them. One can, of course, point out obvious differences between dolphins and sharks, but molecular examination of even more similar organisms can turn up an occasional surprise.

Biologists recognize the relationship between humans and apes, but before DNA became available for study, they considered the two to be in separate families: apes in Pongidae and humans in Homonidae. Chimpanzees and bonobos were considered to be among the Pongidae, more closely related to gorillas and orangutans than to us. DNA analysis has shown, however, that chimps and bonobos are actually more closely related to us than they are to the other apes.

Another reason chimps and humans were once considered to be more distantly related than they are now can be found in the animals' chromosomes, the cellular structures in which DNA happens to be located. Humans have 46 chromosomes, which arrange into 23 pairs, each member of each pair having been donated by each of an individual's parents. In contrast, apes, including chimpanzees, have 24 chromosome pairs. However, the members of human chromosome pair #2 appear to have formed from the end to end fusion of two chromosomes that correspond to the two shortest chromosomes in apes. The ends of chromosomes are chemical structures called telomeres. Human chromosome #2 has telomeres in its center as well as at each end, a completely unique phenomenon in our genome. Moreover, chromosomes have a single more or less central structure called a centromere, which is used when chromosomes reproduce themselves. Human chromosome #2 appears to have a second centromere, albeit a vestigial one. Finally, chromosomes can be stained to show distinct banding patterns. If one places the two shortest ape chromosomes end to end as they are thought to have fused to form Human chromosome #2, one can see banding patters that are almost identical. Moreover, one can show similarities between all other human chromosomes and their corresponding ape chromosomes. The probable explanation for the human/ape chromosome anomaly is that the fusion of the two short chromosomes occurred in an ancestor of humans after our line split off from that of the chimpanzees and bonobos. Similar phenomena have occurred in other organisms.

Admittedly, things would be much easier for biologists if the fusion of the two short ape chromosomes had occurred before humans and chimps separated, and it is unfortunate that we do not have access to the cells of our hominid ancestors to see exactly when the fusion did occur, but the biochemical analysis of the DNA of the respective organisms, and the similarities in their chromosome staining patterns offer pretty strong evidence of our biological relationship.

In summary, there are three major lines of evidence that support evolution. The first is the fossil record, barely touched upon in this chapter. Originally patchy, it has become quite though not totally complete, and every new fossil discovery seems to offer more support for the general, overall concept of evolution having occurred. Secondly, embryological studies have shown how a developing organ appears to start with primitive structures like those of its evolutionary ancestors and modify them into the organs and features that it will use throughout its life. Those organs and features do not simply arise from nothing. Moreover, there is the development of structures that will be abandoned and those that will be maintained only as vestiges or remnants. Finally, there is the biochemical evidence in terms of the fossil genes, DNA similarities among closely related organisms, and similarities in banding patterns on chromosomes. The last category, despite the occasional surprise it has offered, has generally turned out to support the work and conclusions of the pioneers of evolutionary study arrived at from their work on fossils and embryos. It is clear that the science of evolution rests on solid information.

References

Bartlett S (2007) Human ear bones began as reptile jaws. Cosmos online, 16 March 2007. Available at http://www.cosmosmagazine.com/news/1105/humans-ear-bones-began-reptile-jaws

Brusatte SL, Niedzwiedski G, Butler RJ (2011) Footprints pull origin and diversification of dinosaur stem lineage deep into early Triassic. Proc R Soc B 278(1708):1107–1113

Gould SJ (1990) Wonderful life: the burgess shale and the nature of history. W.W. Norton & Co., New York

Chapter 12
Convergence

Every teacher, from those dealing with young children in the early grades to college and university professors, sooner or later comes across a student who is so adamant about some misconception that even when confronted with irrefutable evidence, he or she clings to the fallacy to the point of being psychotic. My worst example was not the one I described in Chap.9, the young woman who thought her Manx cat was half rabbit. It was another one who insisted that bats are birds. After all, they fly! Nothing would persuade her otherwise. I sometimes wonder how she would have reacted had I tried to convince her that sharks are not fish.

When European explorers first came to North America and to South America as well, to some extent they found animals that were largely similar to those they knew or knew of in Europe. The land bridges that existed between Eurasia and North America during periodic ice ages allowed animals on each continent to migrate to the other, and thus the zoologies between the two continents were somewhat alike. The same is true to a lesser extent between North and South America. Thus it was easy for Spanish explorers to see the resemblance between llamas and camels and realize, if they gave it any thought, that the two animals were related. Dutch explorers probably did not have the same reaction upon first seeing Australia, however.

Once the animals of Australia were sufficiently investigated, it was found that many had relatives on the other continents. Australian parrots, crocodiles, monitor lizards, and frogs clearly resemble members of those groupings found elsewhere on the planet. Reptile, fish, and amphibian families on Australia were widely distributed among Gondwanaland continents before Australia became isolated. Birds, because they can fly, were easily distributed around the planet. Only the mammals of Australia as a category appear to be completely unique. Other than bats and dingo dogs, Australian mammals are either marsupials or monotremes, while most mammals across the rest of the planet are placentals. The development of Australian animals in isolation from those elsewhere on the planet accounts for their uniqueness.

B. Marcus, *Evolution That Anyone Can Understand*,
SpringerBriefs in Evolutionary Biology, DOI: 10.1007/978-1-4419-6126-6_12,

Yet, there are some surprises among Australia's mammals. One, the sugar glider, is a small animal with a fur-covered membrane of skin that stretches from almost the end of its forelimb to about the same point on its hind limb. It uses that membrane as an airfoil. It can launch itself out of a tree and glide for as much as 200 feet. The sugar glider, a marsupial, is almost identical to the North American flying squirrel, a rodent. It is more closely related to kangaroos than it is to any squirrel.

There are other types of similar mammals, including the scaly tailed flying squirrels of Africa, another rodent; the Petauridae of Australia, a family of marsupials; and the colugos, a unique family of placental mammals in South-east Asia. All of these have the flap of membrane and the ability to glide for long distances. None of these animals truly fly. There are also gliding lizards, snakes, and frogs. None of these three resemble the gliding mammals, but they have developed a similar habit of launching themselves out of a tree and gliding to another one or to the ground by means of a membrane. Moreover, there were even gliding and flying prehistoric reptiles. Indeed, there are even so-called flying fish, which are actually gliding fish that can launch themselves out of the water and remain airborne for 40–45 seconds and cover more than 1,000 feet.

The similarity in appearance between sugar gliders, flying squirrels, and the other mammals mentioned above is due to a rather common phenomenon in the history of life: convergence. It results when distantly related or unrelated organisms develop similarities in appearance or function, usually as a method to solve a problem they have in common. It is a form of evolution where organisms grow to look alike rather than diverging or growing to look different. The development of gliding in mammals as disparate as rodents, marsupials, and colugos, as well as the independent development of the capability within two subgroups of rodents and marsupials represents convergence. The evolution of flight in two even more disparate groups, bats and birds, does as well.

The logic behind convergent gliding evolution is evident enough. Any arboreal or tree dwelling animal runs the risk of falling, and the smaller the animal happens to be, the less the chance that it will be hurt. For example, a squirrel could fall from the roof of a two-story house without injury. A man falling from the same height very likely would break a leg and a horse would probably explode. Squirrels often leap from tree to tree and sometimes from a tree to the ground. Falling is probably all in a day's work for them. But jumping to the ground or another tree is routine for them, especially if they are being chased by some other animal that is at home in the trees. Most likely, squirrels have a learned sense of how close to the ground or another tree they have to be in order to jump safely, and that distance most likely varies among individuals. Leaping squirrels often spread their limbs, presumably for greater air buoyancy, and those with the loosest skin between their fore and hind limbs would have the best aerodynamics. Thus, any mutation favoring that loose skin would be at a selective advantage. Over time, mutations for more and more loose skin would be favored, and the gliding animals we are describing in this chapter represent the accumulation of such mutations. Similar scenarios can be painted for the development of membranous wings in bats and prehistoric flying reptiles. But it is the gliding membranes of the flying squirrels, sugar gliders, and other gliding mammals described above that show true convergence.

Convergence occurs because similar environmental conditions develop in different locations. If a particular anatomical or behavioral pattern works in a forest in Canada, there is no reason why it would not work in a similar forest in Australia. Thus, if similar mutations occur in different types of animals under similar selective pressures, it stands to reason that both would be successful.

Convergence can be shown in many other cases. One family of marsupials, the Dasyuridae, which also includes the Tasmanian devil, includes a number of small, mouse-like animals. Although they may look like mice, they are more like shrews in behavior in that they are predaceous, eating insects and other small animals, including invasive true mice. The factors leading to their convergence with true mice were probably a matter of their size and habitat preferences.

Often convergence does not result in distantly related or unrelated organisms evolving similarities in appearance. Rather it involves similarities in specific anatomical features, such as teeth. One particularly gross example might be the North American and Eurasian timber wolf and the Australian Tasmanian wolf or tiger. The latter is a marsupial, again more closely related to kangaroos than it is to true wolves, and it is now most probably extinct. However, it did evolve an appearance similar to the timber wolf, and, in particular, it evolved the sharp teeth of predatory animals.

Most people are aware that the large fangs of mammalian predators, technically known as canine teeth, are used to hold prey and to tear flesh from bone. Sometimes, however, it is necessary to cut flesh from bone or to cut flesh into pieces. For this predators use what is called the carnassial teeth. In placental carnivores, it is the last upper premolar tooth and the first lower molar tooth. Wolves use their carnassials for shearing, as do other predatory mammals. The fossil record has shown that this is not a modern adaptation. The Creodonts, an order of extinct placental mammals that lived in the area of 50 million years ago and may have shared a common ancestor with modern carnivores, also had carnassial teeth. However, they were molars farther back in the mouth than is the case with modern carnivores. Likewise, an extinct predator of Australia that has been called the marsupial lion also had carnassial teeth that developed farther back in the jaw. One last example is the Borhyaenidae, a family of extinct South American predatory animals that may have been marsupials or at least closely related to them, also showed the development of carnassial teeth (Carey et al. 1971). The critical point is that all of the predaceous animals we have described, both modern and prehistoric, had the problem of cutting flesh from their prey in order to eat. Simply grabbing mouthfuls and tearing at them would be wasteful in terms of how much would be left behind and the energy required for the activity. All of them, despite their biological distances from one another, solved the problem in similar manners. The carnassial tooth represents a major example of convergent evolution.

Early in this chapter I mentioned that sharks are not fish. I suspect that may have raised an eyebrow or two, but it is technically correct. True fish, known to biologists as bony fish, is the group with which most of us are familiar. It is the class of vertebrates called Osteichthyes, and all of its members have skeletons that are made out of bone. In contrast, sharks and rays are in the class Chondrichthyes; their skeleton is composed of cartilage. It never turns to bone. Both groups most likely descended from a common ancestor well back during the Paleozoic Era, but the branch that led to fish ultimately led to air-breathing vertebrates as well.

Still, the resemblance of sharks to fish cannot be denied, but by the same token, the resemblance of dolphins, porpoises, whales, and manatees to fish cannot be denied either. The body shape of these animals can be described as torpedoform; it is an appropriate shape for animals that live in water. Moreover, the fossil record tells us that during the Mesozoic era, there were reptiles with the same body form, the Ichthyosaurs, a name that means fish lizard.

Animals that live in water and swim freely have to overcome the resistance of a dense medium. They do this in part by having a streamlined shape, one that could be described as hydrodynamic. Many of these animals have to put on a sudden burst of speed to escape a predator or catch prey, and a blocky body would put them at a severe disadvantage. Thus, mutations that led to what we associate as a fish-like shape would tend to be selected. It is, of course, not the only shape that works. Squids, for example are shaped a bit differently, but for the vertebrates it has clearly been the most effective.

One of the most extreme cases of convergence occurs between hummingbirds and a moth known, appropriately enough, as the hummingbird moth, two animals that fall into two entirely separate phyla. Hummingbirds are true birds, phylum Chordata, subphylum Vertebrata, that are quite small. They are capable of hovering over flowers, which they probe with their long, tubular beaks for nectar. By the same token, there are a few species of moths that also hover over flowers and probe them with a long proboscis for nectar. Moths are insects in the phylum Arthropoda and are anatomically much different than birds. Yet hummingbird moths are similar enough in behavior and appearance that people have taken them for hummingbirds, although they are substantially smaller. They even are active during the day, which is characteristic of hummingbirds but not of most moths. However, they come to their appearance from a totally different direction than do hummingbirds. Hummingbird wings evolved from the forelimbs of their ancestors. They are muscular, supported by bone, and covered with feathers. Moth wings evolved off the middle section of their bodies, as did insect wings in general. They are membranous and are supported by veins. Moreover, they are covered with scales.

Curiously, there is a group of nectar-feeding bats that have also evolved the activity of hovering over flowers while they withdraw nectar from them. Known as glossophagine bats, these animals do not have as acute a sense of hearing and are less reliant on sonar than are the insect-eating bats (Castro and Wourms 1993).

Convergence is not unique to animals; it occurs in plants as well. A particular example of this would be the succulent plants of desert environments. In North America, one immediately thinks of cactus as fitting this description, but in Africa and parts of Asia, Euphorbia and even some milkweeds do as well.

Succulents have several adaptations that allow them to conserve water. For one, they generally tend to have fleshy bodies in which they store water, with leaves that are reduced in size or absent altogether. Most of their photosynthesis occurs in their stems, which are often covered with a thick cuticle that also retards water loss.

One last concept to be considered is the development of similarities among organisms without the development of similarities in appearance. This is known as parallel evolution, and it is sometimes hard to distinguish from convergent evolution. Indeed, what some biologists consider to be convergence, others might consider to be parallelism. Consider the long canines of the saber-toothed cat of North America and the saber-toothed marsupial of South America. Were the impressive teeth of these predators a case of convergence or parallelism?

Other examples are not so difficult to define. Embryonic development in placental mammals, for example, is pretty straightforward. A fertilized egg implants in the wall of its mother's uterus and remains there until it has reached the point when it can survive, admittedly often only with extreme care, in the outside world. We frequently use this characteristic as one of the identifying characteristics of mammals. What then of the Atlantic sharpnose shark?

The Atlantic sharpnose shark is a small shark that lives off the east coast of the United States, more typically toward the south. The thing that makes it unusual is that like mammals, much of its embryonic development is at the expense of its mother. Although at first it relies on yolk, as do most sharks, once it has exhausted its supply, it forms a placenta, as do mammals, and obtains its nourishment by way of diffusion from its mother's blood (Howell 1974).

There are other examples of parallel evolution. One concerns the concept of warm-bloodedness in fish and sharks.

Warm-bloodedness is a popular rather than a scientific concept. It is generally thought of as an animal's ability to maintain a constant internal body temperature, or at least one above that of the environment surrounding it. It is generally believed to be characteristic of birds and mammals. In contrast, cold-bloodedness is believed to be the property of an animal's internal body temperature being more or less identical to and determined by the environment around it, and it is thought to be characteristic of all animals other than birds and mammals. Biologists often prefer to use the terms endothermic and ectothermic for warm and cold-bloodedness respectively. Mammals and birds in general are endothermic, because they generate heat by internal biochemical reactions, their metabolism. Some, however, do experience a lowering of body temperature under some conditions, such as when they are asleep or in hibernation. Additionally, some ectothermic animals, like snakes and lizards, can elevate their internal body temperatures behaviorally, such as by basking in the sun.

Fish and sharks are typically ectothermic. However, tuna are capable of maintaining an internal temperature in at least parts of their bodies well above that of the temperature of the water around them, and they do it by trapping heat produced by internal biochemistry. In the case of tuna, the heat is generated in swimming muscles, and it is carried away from those muscles by veins, blood vessels that carry blood laden with waste products away from the muscles for cleansing. Lying next to the veins in tuna are arteries, blood vessels that carry blood rich in nutrient materials and oxygen towards the muscles. Together these vessels actually form a dense network known as the *rete mirabile*, which literally means wonderful net. Tuna generate heat while swimming, and blood circulating through the muscles is carried away via veins. Within the *rete mirabile*, much of the heat in the veins diffuses into the arteries, thus warming the blood within them. The blood then arrives at the muscles pre-warmed, allowing the biochemical reactions that generate heat to occur faster, thus generating even more heat. The overall result is muscle temperature in tunas that is well above that of the waters in which the tunas swim. This allows biochemical reactions like those that bring about muscle contraction to occur rapidly, thus conferring on the tuna speed and endurance. The same mechanism has evolved in lamnid sharks (Marshall 1978), the family that includes makos and great whites.

Regardless of how one chooses to split hairs between convergent and parallel evolution, the point remains that both concepts deal with biological traits that have developed independently in distantly, very distantly, and unrelated organisms. Indeed, since all traits are results of specific biochemical pathways, the traits themselves may have been arrived at from totally separate genetic origins. In all cases, however, the traits occurred in organisms living with similar environmental challenges and subject to similar selective pressures. It should not be surprising that similar solutions to those challenges developed. Dismissing all of that as simply a matter of special creation begs the question of why some of the strategies never occurred in places where they could have survived quite well. For example, why are there no gliding, small arboreal animals in the forests of South America? Why were there saber-toothed predators in the past, according to the fossil record, but no longer? Why did warm-bloodedness occur within only one family each of sharks and fish but in virtually all birds and mammals? The questions can go on and on, but the most easily explained answer, based on hard evidence, is that the phenomena described in this chapter all resulted from changes brought on by natural selection.

References

Carey FG, Teal JM, Kanwisher JWW, Lawson KD, Beckett JS (1971) Warm-bodied fish. Am zool 11:137–143

Castro JI, Wourms JP (1993) Reproduction, placentation, and embryonic development of the Atlantic sharpnose shark, *Rhizoprionodon terraenovae*. J Morphol 218:257–280

Howell DJ (1974) Acoustic behaviors and feeding in Glossophagine bats. J Mammal 55:293–308

Marshall LG (1978) Evolution of the Borhyaenidae, extinct South American predaceous marsupials, University of California Publications in Geological Sciences, vol 117. University of California Press, Berkeley

Chapter 13
Is It Happening Now?

One of the major obstacles to some people when it comes to opening their minds to evolution is the vast amount of time the earth has been in existence and how slowly evolution occurs. Even if punctuated equilibrium were correct in all cases of evolution, the process would still have taken eons and eons, because abrupt change in geologic time is still impossibly slow for humans to observe, given that we typically live for 70–80 years, while "rapid" evolutionary change takes thousands. We do not see evolution happening. Rather, we deduce that it has happened from the evidence we have, just as the police solve crimes they did not see by making deductions from evidence they find.

Other people have accepted evolution, but they believe that it is a phenomenon that occurred in the distant past. It no longer occurs, having somehow ground to a stop. In fact, it continues to occur. It occurs around us, but much too slowly for us to observe unless we look at the right clues.

In Chap. 6, we spoke about so-called superbugs: bacteria that are no longer susceptible to antibiotics and insects that are resistant to pesticides. Both antibiotics and pesticides are largely products of the last century. When they were first put into use, they were remarkably effective at killing the target organisms, and others, at which they were aimed. In Chap. 10 we spoke about the peppered moth and its transition from predominately dark to light and then back to dark coloration. These things have happened on our watch, so to speak. Environmental changes that we or our immediate ancestors brought about have, in turn, brought about changes in organisms with which we share the planet. Such changes are consistent with evolution.

Yet, the argument can be made that the changes described above were in response to human activity. They do not support true Darwinian evolution, which occurs by reason of natural selection and which would have to occur independent of any human involvement. Examples of such phenomena are occurring now; they can be illustrated by what are described as "ring species."

B. Marcus, *Evolution That Anyone Can Understand*,
SpringerBriefs in Evolutionary Biology, DOI: 10.1007/978-1-4419-6126-6_13,

A ring species can be described as a continuous, circular population of organism that can interbreed with neighboring subpopulations along its continuum but not at the very end. If that sounds a bit odd, an example might help clarify things.

Along the west coast of North America, one can find a species of salamander known as *Ensatina eschscholtzii* that occupies woodlands from Baja California in Mexico north to British Columbia, Canada. In California, the salamanders do not exist in the Central Valley, but they do in the mountains surrounding it. All subspecies interbreed with adjacent subspecies, with the exception of two subspecies towards the south of the Central Valley, more or less at the bottom of the ring. Where these meet, they do not seem to interbreed. The subspecies also are colored differently, with the eastern one being blotched bright orange and black, while the western one is more uniformly brown. Given almost everything we have said about species, they would qualify as different species. But they are still members of a common gene pool, and a mutation that occurred in one population, if successfully selected, could hypothetically work its way over many generations up one side of the ring and down the other. Indeed, as one travels north along the east side of the Central Valley, one finds that salamanders change in color until at the north end, the salamanders are intermediate in color between the two subspecies at the southern ends of the ring. This has led at least one investigator to suggest that the ancestral population began at the top of the valley, with descendant populations working their way down each side, differentiating as they expanded their range. By the time the two arms met at the bottom of the valley, the salamanders had differentiated sufficiently that they could no longer interbreed. Biochemical data support this but only to an extent. The northern population is more biochemically diverse than the southern two, which is to be expected. But biochemical data also suggest that the differentiation may not have been smooth and uniform. There may have been instances in the past where subpopulations at ends of the continuum were segregated from and then later reconnected with the main population, which would enhance differentiation. California is an area of earthquakes, and geologic interruption and reconnection of populations is quite possible. The appearances of the animals strongly support the argument of genetic relationship, and the *Ensatina* salamanders may very well be in the process of differentiation at this moment (Dawkins 2004).

Other examples of ring species include the greenish warbler of Eurasia and the crimson rosella parrots of Australia, although the parrots do not appear to be quite so simple a case (Joseph et al. 2008). But again, all of these cases are examples of populations that are genetically diverging from one another, and that is the essence of evolution.

In another instance that may show evolution in progress, a species of walking stick insects (*Timema cristinae*) in the chaparral of southern California, a bit up the coast from Los Angeles, appear to be separating into two subpopulations based on preferences for the plants on which they live. Populations are divided between two species of plants, with each subpopulation preferring the species of plant on which it lives. Thus, the two subpopulations are becoming reproductively isolated from each other. The walking sticks still belong to the same species, but because of the reproductive isolation, gene flow between the subpopulations is restricted, differences are appearing, and divergence appears to be in progress (Nosil et al. 2005).

One last example and we will put this subject to rest. There is a small insect known as the red-shouldered bug; it is a true bug about a half inch long that lives in association with soapberry plants throughout much of the Western Hemisphere. In southern Florida there is also a population that lives on balloon vine. It uses its beak to pierce the seed capsule of the vine to feed on seeds. Between 50 and 60 years ago, a plant from Asia, the goldenrain tree, was introduced as an ornamental. It has since become feral, and red-shouldered bugs have colonized it. The seed capsule of the goldenrain tree is smaller than that of the balloon vine. The bugs that now live on those plants have smaller beaks than those living on balloon vines, although the bugs have not gotten smaller, and they show greater reproductive success (Carroll et al. 2001). These are changes that have occurred in a half-century or so.

One aspect of evolution we have not discussed to this point is the final aspect: extinction. If evolution is change in a species of organisms over time, the eventual ceasing to exist of that species represents a dramatic change. Over the history of life on Planet Earth, five mass extinction events appear to have occurred, based on the fossil record. A sixth appears to be in progress right now. The difference between this and the others, however, is that the previous ones were caused by some kind of natural disaster, such as the comet that caused the extinction of the dinosaurs 65 million years ago. This time, all of the extinctions appear to be driven by the growth and expansion of a single species: us!

To understand our role in the extinction of other organisms, we must briefly consider our own evolution, which will be discussed more thoroughly in the following chapter. Human evolution occurred largely during the Pleistocene Epoch, which began roughly 2.6 million years ago and ended around 12,000 years ago, more or less at the end of the last glacial period, or ice age. It was during this time that we came into our own as a keystone species.

Most evidence suggests that humans originated in Africa. There were a variety of early humans, and on more than one occasion, there were migrations out of Africa, probably across Arabia, and into the Indian subcontinent and Southeast Asia. By 50,000 years ago, our ancestors were modern in all appearances, and populations had left Africa, with some wandering again towards the east. Since then there appears to have been massive extinctions of large, plant-eating animals everywhere humans arrived, more or less at the time they arrived. By 50,000 years ago, humans were armed with spears and shortly thereafter, in geologic time, with slings as well, other than in Australia. This made them effective hunters. Moreover, humans had fire as well, which they may have used to stampede animals or they may have simply lost control of from time to time. The introduction of periodic burning into areas where it had not previously occurred would have had ecological consequences that would have altered habitats, thus making the environment less habitable for many animals.

When humans arrived in Australia, around 55,000 years ago, they found a diversity of marsupials that must have been staggering. Fossils indicate that many were as large as modern, North American deer. There was a flightless bird that weighed over 400 pounds and large reptiles as well. By 50,000 years ago, most of these were gone, and other than some species of kangaroos, there are no longer large marsupials in Australia. A similar extinction occurred around the same time in Southeast Asia. Northern Eurasia once had animals like wooly mammoths, wooly rhinoceroses, mastodons, giant cave bears, and aurochs, the probable ancestors of modern cattle. By 13,000 years ago, these animals were in severe decline. Two thousand years later, more or less coincidental with human arrival in North America, mammoths and mastodons also disappeared there, along with horses, camels, and antelopes. Fossil remains of some of these animals have been found at archeological sites, strongly suggesting interaction with humans. But other animals disappeared from North America around the same time, including saber-toothed cats and giant ground sloths. Around 1,000 years later there was a similar disappearance in South America as humans pushed down there as well. Some unique North American animals persisted in the West Indies until perhaps 4,000 years ago, and New Zealand saw a loss of unique animals, mostly birds, less than 1,000 years ago.

It is unlikely that hunting accounted for the loss of all species, although it probably did some. However, as human populations expanded, there was undoubtedly competition for resources between the humans and the endemic animals, and humans are excellent competitors. Moreover, humans had the means of altering the environment more than any other species. Fire was mentioned above, and in North America, for example, fire was used by Paleo-Indians to clear forests for agriculture and in the existing savanna and grasslands to stampede bison. Oddly, the continent that seemed to have experienced the least impact by humans was Africa, the continent on which humans originated. It is likely that the animals that evolved with our ancestors also evolved strategies to coexist with them.

Over the last few hundred years, as the human population has grown exponentially and human technology has advanced at what can only be described as at a staggering rate, extinctions are occurring as never before. In some cases, deliberate hunting is to blame. Several species of birds, such as the passenger pigeon, the Carolina parakeet, and the great auk in North America were driven to extinction, at least partially by hunting. The introduction of domestic animals and even wild animals for hunting into new environments has caused habitat destruction and competition that some native species could not survive. Even the introduction of plants has had drastic effects by the inadvertent introduction of parasites. The American elm and American chestnut trees have been decimated by fungi that were introduced along with elms and chestnuts from abroad. Whales were hunted mercilessly, and some species no longer exist, while others have been reduced to only remnant populations. Even cod, the fish that essentially supported the colonization of northeastern North America, have been harvested close to the point of population collapse. But more insidiously, the fouling of the environment by the dumping of chemicals into lakes and rivers, the irresponsible use of agricultural chemicals, and the release of acid-forming compounds by the burning of fossil fuels have contributed to environmental pollution and habitat destruction that has badly affected organisms. And now we find that our activities may have altered the climate of the planet. Carbon dioxide buildup in the atmosphere from the burning of fossil fuels traps heat, warming the planet and causing the melting of the polar ice caps. Within the lifetimes of many who I hope will read this book, the North Pole will become ice-free for at least part of the year. When that happens, animals like walruses and polar bears, animals that depend on the ice, will perish. In so doing, they will take their last step in their evolution.

Reference

Carroll SP, Dingle H, Famula TR, Fox CW (2001) Genetic architecture of adaptive differentiation in evolving host races of the soapberry bug, *Jadera haematoloma*. Genetica 112–113:257–272

Dawkings R (2004) The ancestor's tale: a pilgrimage to the dawn of evolution. Houghton Mifflin, Boston

Joseph L, Dolman G, Donnellan S, Saint KM, Berg ML, Bennett ATD (2008) Where and when does a ring start and end? testing the ring-species hypothesis in a species complex of Australian parrots. Proc R Soc B 275:2431–2440

Nosil P, Sadoval CP, Crespi BJ (2005) The evolution of host preference in allopatric versus parapatric populations of *Timema cristinae* walking-sticks. J Evol Biol 119:929–942

Chapter 14
What About Us?

I have a friend who, before he retired, was a science department chairman at the high school of an upper middle class suburb of Rochester, New York. He was also a deacon in his church, but he had a poster hanging in his office that said, "In the beginning God created evolution." I never asked him his views on human evolution, although I assume that he accepted it because he attended number of lectures on the subject. There are, however, a number of people who willingly accept evolution but draw the line when it comes to our own descent. They believe that we are too unique to have come from some other form of life. The evidence suggests otherwise.

Like other forms of life, humans are made up of cells, we start our lives small and get larger by adding cells, we get energy from what we eat and we produce waste materials in which we cannot survive, our basic life functions are governed by DNA, and we die. Regardless of whether or not, there is a supernatural aspect to our existence, there is very much of a biological one, including a fossil record.

On the other hand, there are those who accept the idea of human evolution but have some sizeable misconceptions about what had happened. Among these is that humans evolved from apes, humans represent the pinnacle of evolution, and that some groups within the human species are more highly evolved than others. Additionally, there are people who try to use misconceptions for personal gain. We will discuss some of these later in the chapter, but we will begin with a summary of the timeline that led to us remembering that the actual process was a continuum, not a series of steps, and that the fossil record tells us something about the organisms that lived during our evolution but not necessarily that they were our direct ancestors.

One thing must be stated at the outset: humans did not evolve from monkeys or from apes. Darwin never said that we did; biologists do not believe that we did. Rather, we and the apes evolved from some common ancestor. Most recently it was with chimpanzees and bonobos. Further back in time, there was an ancestor we, along with chimps, shared with gorillas. Still further back in time, we shared a

B. Marcus, *Evolution That Anyone Can Understand*,
SpringerBriefs in Evolutionary Biology, DOI: 10.1007/978-1-4419-6126-6_14,

common ancestor with modern monkeys. These animals share ancestors with us; they are not our ancestors.

Evidence strongly suggests that humans originated in Africa. Most estimates have put the divergence in the lines that led to humans on one hand and chimpanzees and bonobos on the other at around 6–7 million years ago. Much of this was based on calculations using biochemical evidence, such as DNA differences, as fossils from the time were few in number. It is likely that the population of last human/chimp common ancestors was somehow fragmented, with reproductive isolation occurring between the sub population that ultimately led to humans from that which led to chimps and bonobos. At that time, an anthropoid identified as *Sahelanthropus*, who walked on his hind legs, existed. It is unknown whether or not he was directly in our ancestral line, and he may have existed before chimps and humans diverged, but the bipedal locomotion may have served him well if he had to travel across savanna between patches of forest.

Our earliest post divergence ancestors may have been forced to spend more time on the ground than they would have liked to, but it is likely that they retreated to trees whenever they faced a threat. Fossil thigh bones of a creature identified as *Orrorin tugenensis* from around 6 million years ago suggest an animal that walked on its hind limbs but had a head and hands that were more adapted to tree climbing. There is no way of knowing if *Orrorin* was a direct ancestor, but if not, it was probably closely related and similar enough to one.

About a half million years later, lasting for perhaps a million years, the genus *Ardipithecus* was wandering around forests in Africa. A fossil of a female individual from the more recent end of its time span shows a pelvis that was adapted for walking but feet that were still capable of grasping, though less so than a modern chimpanzee's feet. This suggests that *Ardipithecus* spent at least some time on the trees. In addition, the teeth of *Ardipithecus* appear to have been somewhat intermediate between those of modern chimps and humans. The fossil skeleton suggests a body weight that is around 110 pounds, perhaps two thirds that of an average American adult female, although this is something of an apples to oranges comparison. The likelihood is that *Ardipithecus* would have a greater ratio of muscle to fat than modern humans do. However, the brain size of *Ardipithecus* was only about one fifth the size of a modern human's.

Around 4.2 million years ago, a genus that was starting to resemble humans more and apes less emerged. *Australopithecus* lived in Africa for 2 million years, developed into several species, including a probable direct ancestor of ours, and walked upright. It is unknown which species gave rise to the genus *Homo*, our own, but *Australopithecus*, although probably an agile climber, appeared to have spent more time on the ground than in the trees. Evidence to its being upright includes a set of footprints in petrified volcanic ash that have been dated to around 3.6 million years ago. The footprints were found near Laetoli in northern Tanzania and appear to show two individuals of different size walking side by side. Whether it was a male and female or an adult and juvenile we do not know, but the feet were more human than ape like in appearance. It is postulated that the pair crossed the ash not long after it had settled.

Among the most known fossils of *Australopithecus* is one that has come to be known as "Lucy." Lucy, as the name suggests, was a female *Australopithecus* from around 3.2 million years ago. She stood a few inches less than four feet tall, but the skeletal remains, about 40% of a complete skeleton, show enough to indicate that she stood erect and moved about on two feet, not four. In general, *Australopithecus* had a brain that was about four-tenths the size of ours, indicating that upright posture preceded the development of our large, for our body size, brain.

Upright posture and walking around on two feet, being bipedal, along with a large brain size are considered to be two of the most characteristic features of humans. Fossils show that bipedal locomotion actually goes back a long way, perhaps even previously to the human and chimp/bonobo divergence. If that is indeed the case, chimpanzees might actually have readapted to a less erect posture that is preferable for a more arboreal life. In other words, their ancestors may have returned to the forests after experiencing life on the ground.

Nobody can say for certain how bipedal posture came about. There are numerous hypotheses about being able to see better in tall grass, having less body surface exposed to the direct rays of the sun to prevent overheating, and freeing the hands in order to better carry food back to a mate and offspring. The critical point is that the mutations that produced genes for upright posture in our ancestors were beneficial in the environment in which they occurred; consequently, they provided a selective advantage to the individuals that carried them. Those individuals reproduced more successfully and left behind more offspring. As our evolution continued, those mutations continued to serve our ancestors well.

As mentioned, there were several species of *Australopithecus*, some of which coexisted. Additionally, there were other genera of what might be called ape-men extant in Africa at the time of *Australopithecus*, and fossils suggest that tool use may have been occurring toward the end of their time on the planet as well. Earliest tools were probably little more than rocks, boughs, or bones that were found and put to use, but by the time the first members of the genus *Homo* appeared, our ancestors were probably fashioning tools.

Earliest members of our genus evolved from *Australopithecus* and were extant by around 2.3 million years ago. This genus was more like modern humans in terms of its teeth, its hands, and its larger brain size. Among the first of members of this genus was *Homo habilis*, once thought to actually be the first member of genus *Homo* on the planet and also once thought to be the first being to use stone tools. Both of those ideas are now disputed, as is the notion that he was definitely a direct ancestor of ours. *H. habilis* was fairly small, only around four feet tall, and his brain size was slightly less than half of ours. But he survived as a species for almost a million years, sharing Africa with a number of other species of genus *Homo*, and he coexisted with some pretty fierce predators, without having been wiped out by them. Clearly he had something going for himself.

Among the species of *Homo* that coexisted with *H. habilis*, for at least part of their tenure on the planet, were *Homo ergaster*, from about 2.5 to perhaps 1.7 million years ago, and *Homo erectus*, from about 1.8 million years ago to

300,000 years ago. Whether or not *H. habilis* was the ancestor of both of these species is under discussion, but the possibility exists that it was. Moreover, *H. ergaster* may be intermediate between *H. habilis* and *H. erectus*, or the two may have been sister species. But it was during the existence of these species that the first known members of genus *Homo* wandered out of Africa. Fossils of *H. erectus* have been discovered in the nation of Georgia, a former Soviet republic on the Black Sea north of Turkey and south of Russia, in a region known as Dmanisi. The earliest fossil find dates to around 1.85 million years ago, which raises questions. Among them, did *H. erectus* originate in Africa earlier than we think and migrate out of Africa, or did some other species of *Homo*, say *H. ergaster*, migrate to Eurasia and give rise to *H. erectus*? And are the fossils that are now identified as *H. erectus* in Africa truly *H. erectus*, or are they members of closely related but separate species? Whatever the answers happen to be, *H. erectus* was a force to be reckoned with.

Standing at around six feet tall, a male *H. erectus* was a well-muscled being with a brain size about 850 cubic centimeters (cc) early in their history. Later specimens may have had brains as large as 1,100 cc, which approaches to that of modern humans and suggests that he may have used language. Moreover, he had the use of fire as well as stone tools, he most likely had a social structure that was more complex than earlier species of *Homo*, and his appearance was more human and less ape like, and females were closer in size to males. These characteristics make *H. erectus* more like us than like apes.

H. erectus was perhaps the most widespread of all members of our genus, other than us. He managed to make it throughout Europe and Asia, as well as Africa. References to Java man and Peking man actually refer to Asian discoveries of *H. erectus*.

As mentioned previously, it is impossible to categorically state that the species described in this chapter represent a direct line from the last human/chimpanzee common ancestor to modern man. *H. erectus* may have been an ancestor; he may not have been, but the fossil record is replete with evidence of his and other species' existence. Our ancestors were among them. One of the biggest criticisms one hears about human biological history is that there are gaps in the fossil record; there is no so-called "missing link," an intermediate species between humans and apes. The criticism probably dates back more than a century, and it is invalid. All of the species discussed in this chapter are represented by fossils. More are being discovered all of the time. At risk of being overly repetitive, we can state that the fossil record has become rich enough to demonstrate that there has been a wide diversity of species within genus *Homo* and its immediate predecessors, not simply the ones mentioned here. Indeed, to describe them all would take a volume much larger and greater in detail than this one. Moreover, one must remember that the probability of any single individual being fossilized is remote. Before our ancestors or their relatives began to seriously inter their dead, deceased members of their communities, such as they were, most likely were abandoned and left for scavengers. Any bone fragments left behind by carrion-eating animals most likely decayed fairly rapid. Consequently, the fact that the record of humanoid fossils is as rich as it is suggests a diversity of species that Darwin and other pioneers in the study of evolution most probably could never have imagined. In addition, the punctuated equilibrium model of evolution postulates that changes in species can occur quite rapidly in terms of geologic time. With that being the case, it is only natural that gaps in a fossil record would exist. Even so, much of the human evolution can be inferred from DNA, and much more will be learned as more studies are conducted and more fossils are discovered. Finally, every time an intermediate fossil is discovered, one still hears criticism that there are no intermediates between it and some previous or later fossil. In other words, there is no missing link. That argument is becoming tiresome.

The earliest modern human beings, *Homo sapiens*, first appeared around a half million years ago. Sometimes referred to as "dawn man," these people may have comprised a variety of human types, possibly even a number of species. Whether they emerged from *H. erectus* or a sister or descendant species, with a cranial capacity of 1,300 cc, almost the equivalent of our own on average, they represent a quantum change. Modern *H. sapiens* may have descended from archaic ancestors as early as 100,000 years later, but the fossil record shows that archaic forms coexisted with modern ones until perhaps less than 100 millennia ago. The archaics were probably more in keeping with what we think of as "cave men" in appearance than we are. They were probably more muscular and more resembling of *H. erectus*, and we may not be their only descendants (Dawkins 2005). Our sister species, *Homo neanderthalensis*, Neanderthal man, may be the only species of humans that did not originate in Africa.

Neanderthal man is not an ancestor of modern humans; he was a contemporary until around 30,000 years ago, when he went extinct. Nobody knows why Neanderthals died off, although a number of hypotheses have been suggested, including that he was a victim of genocide by contemporary *H. sapiens*. Neanderthal's disappearance coincided with a rapid climate change in Europe, where they lived. It is possible that he did not adapt as well as *H. sapiens* did and was simply out competed in a new environment.

Neanderthal man existed for more than 300,000 years. He was not the dumb brute of popular perception. Rather he was as technically proficient and socially advanced as the *H. sapiens* with whom he shared Europe and western Asia. He had fire, he made tools, he lived in a societal setting, and it is probable that he had some sense of religion. Fossil finds in what appear to be burial sites suggest that the Neanderthal dead were laid to rest with care and ceremony rather than simply abandoned to scavengers.

Fossil evidence suggests that Neanderthals originated outside of Africa, probably from a population of *Homo* that had migrated out. Modern humans originated in Africa from a population of the same ancestral species that remained there, and it was not until *H. sapiens* had fully evolved before modern humans migrated out.

Scientists in Germany have succeeded in isolating and sequencing Neanderthal DNA from fossil bone, and they have shown that it is very similar to our own, indicating that Neanderthals and modern humans were very closely related. In addition, analysis shows that interbreeding occurred between Neanderthals and modern humans, and that those of us with European, Middle Eastern, or Asian ancestry carry some Neanderthal genes in our genomes. Only those with no ancestry north of Sub-Saharan Africa do not (Green and others 2010).

Other dead end chains in human evolution within the most recent 100 millennia exist. One case involves 40,000 year-old fragments of bone and tooth found in the Denisova Cave in mountains in Siberia in central Asia. DNA analysis suggests that the individual from whom those fragments came was neither modern human nor Neanderthal. Even more striking is a tiny human from the Island of Flores in Indonesia. Standing at around three feet tall, with a weight of perhaps 60 pounds and a brain about one third the size of modern humans, this individual has become known as *Homo floresiensis* and because of its diminutive size has been nicknamed The Hobbit.

Skeletal remains of The Hobbit range from around 95,000 to 13,000 years ago, just yesterday in terms of the geologic time scale. Exactly how it got to Flores is not known, but The Hobbit's archaic bone structure is suggestive of *H. erectus*, and it has been postulated that this is the species from which it evolved. The thinking is that The Hobbit's ancestors got to Flores and over the passage of time evolved into The Hobbit. It is not unusual for large animals on islands to evolve to a smaller size. Indeed, The Hobbit hunted a stunted elephant on Flores. Exactly why The Hobbit went extinct is also unknown, but it has been suggested that it was eliminated by a volcanic eruption on Flores somewhere around 12,000 years ago. It is also possible that modern humans got to the island and out competed their tiny ancestor (National Geographic Society 2011).

Modern man, the only surviving species in genus *Homo* emerged around 100,000 years ago in Africa. Molecular biological studies suggest that it might be something of a marvel that the pioneer members of our species managed to survive. Our ancestors passed through what are called genetic bottlenecks during our biological history, during which population number was reduced and survival tenuous. One appears to have occurred sometime around 70,000 years ago, when a volcanic super eruption occurred on the island of Sumatra in Indonesia. So much debris was ejected into the atmosphere that sunlight was markedly reduced, and a volcanic winter that may have lasted for 10 years resulted. During that time, the human population in Africa dwindled. A similar but smaller bottleneck may have occurred when the ancestors of the first non African humans migrated out of Africa. These individuals were few enough in number that they could easily have been wiped out by some event like disease or drought. Other possible bottlenecks may have occurred when our ancestors had to adapt to upright posture, which brought about skeletal changes making what amounts to premature birth essential. If humans were born at the same stage of development as chimpanzees, for example, our heads would be too large to pass through the birth canal. In essence, genetic bottleneck is the biological equivalent of passing through the eye of a needle. We are lucky that our ancestors made it. Parenthetically, the idea that humans represent some evolutionary ideal is disputed by our being born prematurely but not so prematurely that it does not put a serious strain on our mothers. Childbirth was once a major cause of death among women. (Fig. 14.1)

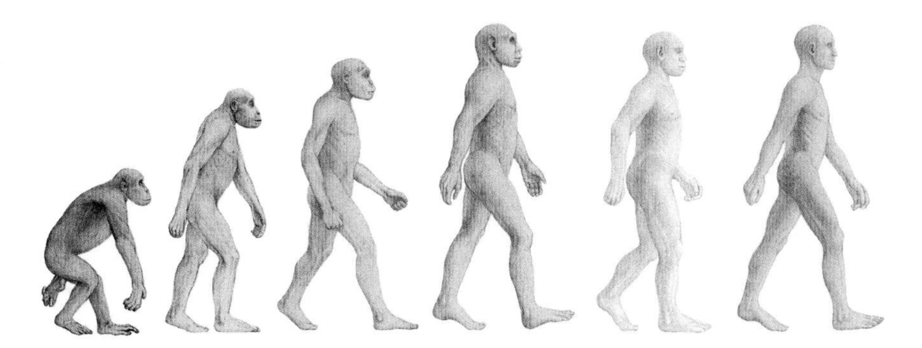

Fig. 14.1 Parade of human evolution, from a common ancestor of humans and chimpanzees to modern *H. sapiens*. Courtesy of Seven Continents History. Reproduced with permission

The idea of genetic bottlenecks brings up a fascinating topic in the realm of human evolution; there might actually have been an Adam and Eve, although their story is entirely different from the one told in Genesis. To understand it, we must digress a bit.

The idea that every one of us is the result of the union of two cells, collectively our mothers' ova and our fathers' sperm, is hopefully common knowledge. Moreover, that each of our parents contributed 50% of our individual DNA should also be commonly known. However, the latter sentence is not entirely correct, and Our mothers contribute more than simply DNA, while our fathers do not. Each sperm cell is little more than a cell nucleus, where DNA is located. Whatever other components it has are discarded when the sperm unites with an ovum. In contrast, the ovum is a nearly complete cell, with a full complement of the cellular components that are found outside the nucleus. Among those components is a collection of small structures known as mitochondria, mentioned earlier, and it is in the mitochondria that food molecules ultimately react with oxygen to release energy. Clearly our cells and by extension ourselves, could not survive without mitochondria, which we get only from our mothers, and mitochondria, which reproduce separately from the cell that contains them, have their own DNA. Moreover, since mitochondria reproduce by copying themselves, the only way for mitochondrial DNA, in theory anyway, to change is by mutation, that is by an error in copying. Such errors do occur at a predictable rate. Using that data, molecular biologists can calculate the rate of change mitochondrial DNA undergoes and in so doing have been able to date specific mutations and pinpoint times of divergence. It also allows them to trace our matrilineal genealogy, and in so doing, they have come up with an ancestral lineage that suggests that all human mitochondria are descendant from those of a theoretical single female individual who lived in Africa around 200,000 years ago (Rice University 2010). Nicknamed "Mitochondrial Eve," the woman who carried these original mitochondria, if indeed she existed, represents as tight a genetic bottleneck as one could imagine: a single individual. Had anything happened to her before she reproduced, the human race now would be entirely different from what it is, if it existed at all.

Along similar lines, patrilineal ancestry can be traced using the DNA on the Y chromosome. Remembering back to Chap. 10, where sex determination was described, only males have a Y chromosome, which they inherit from their fathers. Y chromosomes normally do not exchange genetic material with their partner X chromosomes. Consequently, the techniques used in tracing mitochondrial DNA have been used to trace Y chromosome DNA, and the results suggest that a Y-chromosome Adam lived in Africa, and a recent study placed his existence at around 140,000 years ago (Cruciani et al. 2011). Previous studies had him living more recently. But whenever he existed, if, in fact, he did exist, he too represents a major genetic bottleneck that could have spelled the end of our species had he not reproduced.

The existence of Mitochondrial Eve and Y-chromosome Adam is not a topic of universal agreement among biologists, and from my perspective it is something best argued among molecular biologists. But research into the issues is ongoing, and eventually the correct answer will be available to us. We will have to be patient to see how this plays out.

Unfortunately, some so-called proponents of human evolution are as misinformed about the subject as are its detractors. Earlier we mentioned that our individual entry into the world does not put us at the pinnacle of evolution. Indeed, it has historically been a cause of problems. Yet some people still insist that we are as close to perfect as evolution or creation could come. Indeed, one of our premier characteristics, our upright posture, is the source of some of our greatest ills. Our posture puts a substantial strain on our lower backs, our hips, and our knees. Consequently, arthritis often sets into these areas causing ongoing pain. During our distant ancestry, when our life span rarely exceeded 35 years, chronic pain was not much of an issue. Today, however, when so many of us routinely reach more than twice that age, many of us also suffer from chronic back, hip, and knee pain as well.

There are others who use evolutionary misinformation for personal profit, whether out of greed or ignorance. One case in point is the contention among some people that one's diet should be dictated by one's blood type. The argument is that the different blood types appeared during different stages of human evolution and that one's diet should be adjusted accordingly. For example, proponents of the blood type–diet relationship hold that type O is the oldest human blood type; it dates back to the hunter-gatherer days of our ancestors. Consequently, people with type O blood should eat a diet rich in protein and fat, as our hunter-gatherer ancestors supposedly did. In contrast, type A blood appeared when our ancestors began farming and cultivated plants. Thus, people with type A blood should concentrate more on plant-based foods. The problem with that is that type A probably has been around longer than O. Indeed, biochemically it would make more sense that O is the mutation and A is the norm. Type A is the common blood type in chimpanzees, our closest relative. One might reasonably infer our most recent common ancestor had type A blood. Should someone with type A blood eat like a chimpanzee? Moreover, it is well within the realm of probability that someone with type O blood could have one or two type A parents. Is it reasonable that his diet should be more like a total stranger's who happens to have the same blood type than it should be like the people with whom he shares his DNA? Not everyone who speaks for evolution speaks accurately.

More insidious than the inaccuracy described in the previous paragraph is the idea that some people are more recently evolved than others and are therefore somehow superior. Evolution has been used historically and inaccurately to support blatant racism. People of African descent have been the greatest victims of this kind of prejudicial thinking, which is absurd given that all contemporary humans descended from Africans. Moreover, those of us of European or Asian ancestry descended from a small group of people who emigrated from Africa. There was nothing superior or smarter about them; they simply migrated possibly following herds of animals that they hunted. Even today, there is more genetic diversity among African people than there is among all non-Africans combined, rendering the concept race being simply a matter of skin color scientifically ridiculous.

At this point it would be reasonable to ask if humans are still evolving. It is really hard to say. Even if our evolution has been entirely a matter of punctuated equilibrium, the changes that occur still do so slowly in terms of the human life span. We cannot observe change happening. We can point out change that has occurred in a relatively short time. We know from measurements made on ancient skulls that our brains are smaller than those of our somewhat recent ancestors. The reduction does not necessarily mean that we are any less intelligent than our ancestors; it may be only a matter of our simply needing less of our brains to work on survival. Few of us have to worry about predators or raiders from another tribe, as our ancestors did. Furthermore, many of us tend to be skilled in highly specific areas rather than being generally skilled as were our ancestors. And our smaller brains give us the luxury of not having to support and provide for an energy hungry larger central nervous system. However, it will be only with the wisdom of hindsight that we will know whether or not we are still evolving. It is a question that will be answered by our descendants.

References

Cruciani F, Trombetta B, Massala A, Destro-Bisol G, Sellitto D, Scozzari R (2011) A revised root for the human Y chromosomal phylogenetic tree: the origin of patrilineal diversity in Africa. Am J Hum Genet 88:814–818. http://www.cell.com/AJHG/fulltext/S0002-9297%2811%2900164-9. Accessed 16 June 2011

Dawkins R (2005) The Ancestor's tale: a pilgrimage to the dawn of evolution. Houghton-Mifflin Co, Boston

Green RE et al (2010) A draft sequence of the Neandertal genome. Science 328:710–722

National Geographic Society (2011) The people time forgot: flores find. National Geographic. http://ngm.nationalgeographic.com/features/world/asia/georgia/flores-hominids-text. Accessed 14 June 2011

Rice University (2010) 'Mitochondrial eve': Mother of all humans lived 200,000 years ago. Sci Daily. http://www.sciencedaily.com-/releases/2010/08/100817122405.htm. Accessed 15 June 2011

Index

B. Marcus, *Evolution That Anyone Can Understand*,
SpringerBriefs in Evolutionary Biology, DOI: 10.1007/978-1-4419-6126-6,